T0222168

Signal Processing for Solar Array Monitoring, Fault Detection, and Optimization

Synthesis Lectures on Power Electronics

Editor
Jerry Hudgins, *University of Nebraska, Lincoln*

Synthesis Lectures on Power Electronics will publish 50- to 100-page publications on topics related to power electronics, ancillary components, packaging and integration, electric machines and their drive systems, as well as related subjects such as EMI and power quality. Each lecture develops a particular topic with the requisite introductory material and progresses to more advanced subject matter such that a comprehensive body of knowledge is encompassed. Simulation and modeling techniques and examples are included where applicable. The authors selected to write the lectures are leading experts on each subject who have extensive backgrounds in the theory, design, and implementation of power electronics, and electric machines and drives.
The series is designed to meet the demands of modern engineers, technologists, and engineering managers who face the increased electrification and proliferation of power processing systems into all aspects of electrical engineering applications and must learn to design, incorporate, or maintain these systems.

Signal Processing for Solar Array Monitoring, Fault Detection, and Optimization
Henry Braun, Santoshi T. Buddha, Venkatachalam Krishnan, Cihan Tepedelenlioglu, Andreas Spanias, Toru Takehara, Ted Yeider, Mahesh Banavar, Shinichi Takada
2012

The Smart Grid: Adapting the Power System to New Challenges
Math H.J. Bollen
2011

Digital Control in Power Electronics
Simone Buso and Paolo Mattavelli
2006

Power Electronics for Modern Wind Turbines
Frede Blaabjerg and Zhe Chen
2006

Signal Processing for Solar Array Monitoring, Fault Detection, and Optimization

Henry Braun, Santoshi T. Buddha, Venkatachalam Krishnan, Cihan Tepedelenlioglu, Andreas Spanias, Toru Takehara, Ted Yeider, Mahesh Banavar, Shinichi Takada

ISBN: 978-3-031-01369-0   paperback
ISBN: 978-3-031-02497-9   ebook

DOI 10.1007/978-3-031-02497-9

A Publication in the Springer series
SYNTHESIS LECTURES ON POWER ELECTRONICS

Lecture #4
Series Editor: Jerry Hudgins, *University of Nebraska, Lincoln*
Series ISSN
Synthesis Lectures on Power Electronics
Print 1931-9525   Electronic 1931-9533

Signal Processing for Solar Array Monitoring, Fault Detection, and Optimization

Henry Braun, Santoshi T. Buddha, Venkatachalam Krishnan, Cihan Tepedelenlioglu, Andreas Spanias, Toru Takehara, Ted Yeider, Mahesh Banavar, Shinichi Takada

SYNTHESIS LECTURES ON POWER ELECTRONICS #4

ABSTRACT

Although the solar energy industry has experienced rapid growth recently, high-level management of photovoltaic (PV) arrays has remained an open problem. As sensing and monitoring technology continues to improve, there is an opportunity to deploy sensors in PV arrays in order to improve their management. In this book, we examine the potential role of sensing and monitoring technology in a PV context, focusing on the areas of fault detection, topology optimization, and performance evaluation/data visualization. First, several types of commonly occurring PV array faults are considered and detection algorithms are described. Next, the potential for dynamic optimization of an array's topology is discussed, with a focus on mitigation of fault conditions and optimization of power output under non-fault conditions. Finally, monitoring system design considerations such as type and accuracy of measurements, sampling rate, and communication protocols are considered. It is our hope that the benefits of monitoring presented here will be sufficient to offset the small additional cost of a sensing system, and that such systems will become common in the near future.

KEYWORDS

photovoltaic systems, electrical fault detection, condition monitoring, circuit optimization, wireless sensor networks

Contents

CHAPTER 1

Introduction

Photovoltaic (PV) electrical power generation is an active and growing area of academic research and industry. In 2009, the most recent year for which such statistics are available, a record high of nearly 1.3 peak Gigawatts of PV generation capacity were manufactured and shipped by U.S. companies, a nearly 30% increase from 2008 [1]. Although rapid and continuous progress has been made in increasing efficiency and reducing cost of PV electric power generation, array level management of PV installations remains much the same as in previous decades. The family of technologies and protocols collectively known as the smart grid offer an opportunity to change this: with increased monitoring and communication among PV array components, significant improvements in overall array power production may be achieved.

As with other renewable energy sources, progress in PV is motivated primarily by a desire to reduce fossil fuel consumption. Increasing the share of renewables such as PV in electrical generation has clear environmental and political benefits: PV modules produce no greenhouse gasses during operation and relatively little during manufacturing, and do not require significant resources which must be imported from other countries. As such, governments around the world have subsidized the construction of PV arrays and funded research into PV technology.

Despite these advantages, however, PV technology faces several barriers which prevent it from being widely deployed. The most notable barrier is cost: in the US as of 2010, the average levelized cost of energy (LCOE) for PV electricity was \$211/MWh, while the LCOE of coal generation was only \$95/MWh [2]. PV has achieved cost parity with conventional energy only for a few special cases, such as very remote locations where fuel shipping costs are extremely high. The other barrier to widespread deployment of PV is its intermittence: PV array output is dependent on the weather, and with no effective way of storing energy, PV plants may cause stability and reliability issues for the electrical grid. Any technology which lowers cost or improves array reliability will accelerate the acceptance of PV into the mainstream of power generation. By dynamically optimizing array electrical configuration and by replacing current ad-hoc methods of monitoring and fault detection, the efficiency of PV arrays may be increased while decreasing the net cost per kWh of electricity from PV sources.

This book describes several techniques based on information processing which may be used to improve the operation and management of PV arrays. The specific case of a "smart PV module" which reports its state of operation and is capable of dynamically switching its electrical configuration is used. In general, two main problems are considered. First, the presence of faults in a PV system may be detected and more quickly mitigated by an automatic system. Second, operation in normal (no fault) conditions may be dynamically optimized, e.g., by altering the array topology. Topics and background information are presented in such a way as to be useful for readers with a background in either renewable energy or signal processing.

Automated fault detection is the first potential application discussed for signal processing in PV. Current methods of fault detection are not rigorous, are slow, and rely on human operators. In [3], a faulty array which operated for roughly 1 month before being repaired is described. In [4] mean time to repair (MTTR) of several PV installations was calculated; all installations showed a MTTR of at least 19 days. The authors of [5], on the other hand, identify a MTTR of 3.3 days for a large-scale PV system; this is noted as being a very short time. With automated fault detection and mitigation, time to repair could be dramatically reduced, in some cases to under 1 h.

The next topic considered is dynamic optimization of PV array topology in order to increase operating efficiency. In general, this approach allows the optimal operating voltage of the array to be moved to one of a small number of values. This operating voltage in turn affects the magnitude of resistive losses in connection cables and slightly alters inverter operating efficiency, among other possible effects. The choice of array topology also helps to determine the magnitude of losses due to mismatch between PV modules. Mismatch occurs when PV modules' electrical parameters differ from one another; this can occur due to aging and failure of modules, partial shading of the array, build-up of dirt on module surfaces, or imperfections in manufacturing. In cases where a faulty module exists, dynamically switching array topology has the potential to dramatically increase array output until a full repair.

The book is organized as follows. In Chapter 2, a brief overview of the underlying physics and high-level electrical behavior of PV cells, modules, and arrays is presented. Two distinct but related PV module performance models, developed at Sandia National Labs [3] and University of Wisconsin-Madison [6], are described and compared. The high-level operation of a PV array is also discussed.

Chapter 3 discusses several of the most common faults and causes of reduced performance which occur in PV arrays. Their effect on power output and array electrical behavior is discussed, as are any safety issues posed by the presence of a fault. The effects of mismatched, soiled, and shaded PV modules are described first. These conditions reduce the performance of a PV array but may be considered a normally operating system. There is little or no current practice to mitigate these conditions once they occur. Next, ground faults, DC arcs, high-resistance connections, and islanding are described. Unlike mismatches, These conditions may truly be considered faults: they can cause a dramatic drop in power output, damage to equipment, or safety hazards.

Chapter 4 presents statistical signal processing-based techniques for determining the presence of a fault. Classical distance-based tests are presented first, followed by more robust modifications which account for the non-independence of measurements from within a PV array. Finally, machine learning-based methods of fault detection are described.

Current methods for optimally sizing PV arrays are discussed in Chapter 5, and the need for a dynamically reconfigurable array is demonstrated. Existing approaches to constructing reconfigurable

PV arrays are then discussed followed by a brief description of the requirements for a reconfigurable PV array design.

Finally, in Chapter 6, a brief overview of monitoring aspects relevant to information processing of PV arrays is given. The need for monitoring on the DC and AC side of the inverter is described. Monitoring systems currently used in literature are reviewed. The various considerations for a monitoring such as placement of sensors, communication systems, and user interfaces are discussed.

CHAPTER 2

Overview of Photovoltaics

The photovoltaic cell is the fundamental power conversion unit of a PV system [7] and is the component that produces electricity from solar energy. Although a single cell is capable of generating significant current, it operates at an insufficient voltage for typical applications. To obtain a higher voltage, cells are connected in series and encapsulated into a PV module. These modules show similar electrical behavior to individual cells. Similarly, modules are connected in series and parallel to form a photovoltaic array. The arrays generate direct current (DC) power which is converted to alternating current (AC) power using inverters. The physics of photovoltaic cells and their electrical behavior are discussed in Section 2.1 and some commonly used models to analyze their performance are described in Section 2.2. The electrical behavior of photovoltaic arrays is discussed in Section 2.3.

2.1 OPERATION OF A PHOTOVOLTAIC CELL

Photovoltaic cell operation is based on the ability of a semiconductor to convert sunlight into electricity through the photovoltaic effect [7]. When light is incident on the solar cell, incoming photons can either be reflected, absorbed, or passed through it. Only the absorbed photons contribute to the generation of electricity. For a photon to be absorbed, its energy must be greater than the band gap of the solar cell, which is the difference between the energy levels of the valence band and conduction band in the cell. The absorbed photons generate pairs of mobile charged carriers (electrons and holes) which are then separated by the device structure (such as a p-n junction) and produce electrical current. A variety of materials facilitate the photovoltaic effect. In practice, semiconductor materials in the form of p-n junctions are mostly used to manufacture solar cells. To understand the operation of a solar cell, it is essential to know the functioning of p-n junctions.

Consider a p-n junction in a semiconductor. There is an electron surplus in the n-type semiconductor and a hole surplus in the p-type semiconductor. At the junction of the two semiconductors, the electrons from the n region near the interface diffuse into the p side. This leaves behind a layer of positively charged ions in the n region. In a similar fashion, holes diffuse in the opposite direction leaving behind a layer of negatively charged ions in the p region. The resulting junction region is devoid of mobile charge carriers. The positively and negatively charged ions (dopant atoms) present in the junction region result in a potential barrier which restricts any further flow of electrons and holes (as shown in Figure 2.1). This potential barrier is known as the depletion region. The resultant electric field in the junction pulls electrons and holes in opposite directions [7]. Therefore, current flow through the junction requires a voltage bias.

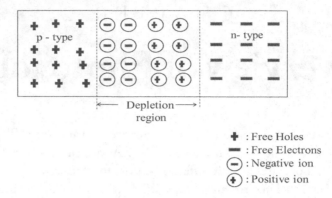

Figure 2.1: The p-n junction barrier formation.

When an external bias is applied to the junction (Figure 2.2), for instance a negative voltage to the n-type material and a positive voltage to the p-type material, the negative potential on the n-type material repels electrons in the n-type material and drives them towards the junction. Similarly, the positive potential on the p-type material drives the holes towards the junction [8]. This reduces the height of the potential barrier. Consequently, there is a free motion of charges across the junction resulting in a dramatic increase in the current through the p-n junction. This is known as the *forward bias* situation.

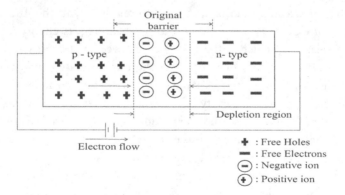

Figure 2.2: Forward biased p-n junction.

When *reverse biased*, the p-type material is biased at a lower voltage than the n-type material (Figure 2.3). The electrons in the n-type material are drawn towards the positive terminal and holes in the p-type material towards the negative terminal. Therefore, the majority charge carriers are pulled away from the junction. This results in an increase of the number of positively and negatively charged ions (dopant atoms), widening the depletion region [9]. Thus, a continuous motion of charges is not established because of the high resistance of the junction [8] and the junction is said to be reverse biased. However, the junction appears to be forward biased and provides low resistance to the minority carriers (electrons in p-type and holes in n-type regions). These minority carriers result in a minority current flow. This current, known as the reverse saturation current (I_0), is much smaller in magnitude than the current generated under forward bias. In the presence of an external source of energy such as light, heat, etc., the electron hole pairs generate minority carriers which contribute significantly to current flow across the junction.

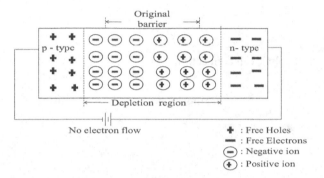

Figure 2.3: Reverse biased p-n junction.

The I-V characteristic of a p-n junction diode is given by the Shockley equation [7]

$$I_D = I_0 \left[\exp \left(\frac{q V_D}{n k T_{cell}} \right) - 1 \right], \tag{2.1}$$

where I_D is the current generated by the diode, V_D is the voltage across the diode, I_0 is the reverse saturation current of the diode (usually on the order of 10^{-10} A), $q = 1.602 \times 10^{-19}$ Coulombs is the electron charge, T_{cell} is the cell temperature in Kelvin, n is the diode ideality factor (dimensionless), and $k = 1.38 \times 10^{-23}$ J/K is the Boltzmann constant.

A p-n junction can be made to operate as a photovoltaic cell [8] (Figure 2.4). The p-n junction responds to incident light and generates electric current. The influence of arriving photons produces a minority current effect [8]. These photons generate free electron-hole carriers which are attracted towards the junction. The electron and hole charges travel in opposite directions and set the direction of the photovoltaic current, as shown in Figure 2.5 [8]. The electron flow in the circuit (shown in Figure 2.5) is from n-type silicon to p-type silicon [8]. The generated current varies with the light intensity.

Ideally, the solar cell is electrically equivalent to a current source in parallel with a diode (Figure 2.6) [10, 11]. This circuit can also be used to represent a PV module or array [12]. The following equations that describe the electrical behavior of solar cells can also be applied to solar modules.

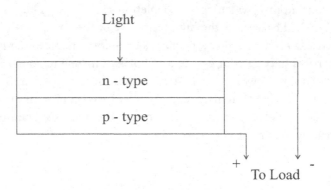

Figure 2.4: The photovoltaic cell connection.

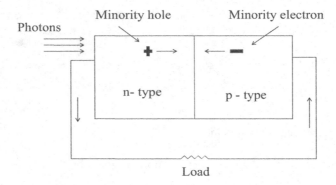

Figure 2.5: Operation of photovoltaic cell.

The sign convention used for current and voltage in photovoltaics is such that the photocurrent is always positive. As shown in Figure 2.6, the light generated current, also known as photocurrent, is represented as I_L, the diode current as I_D, the net current and terminal voltage of solar cell as I_{cell} and V_{cell}, respectively. The net current I_{cell} available from the solar cell is given as

$$I_{cell} = I_L - I_D. \tag{2.2}$$

Substituting I_D from Equation (2.1) into Equation (2.2) [11],

$$I_{cell} = I_L - I_0 \left[\exp\left(\frac{q V_{cell}}{nkT} \right) - 1 \right]. \tag{2.3}$$

Light-generated current I_L increases linearly with solar irradiation. The smaller the diode current I_D, the more current is delivered by the solar cell. The ideality factor n of a diode is a measure of

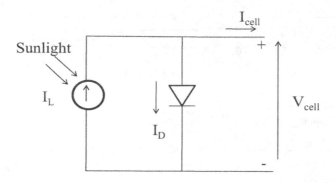

Figure 2.6: Equivalent circuit of an ideal photovoltaic cell.

how closely the diode follows the ideal diode equation [10, 11]. Typically, it takes values in between 1 and 3 [11]. The value $n = 1$ represents the ideal behavior of diode, while values $n > 1$ correspond to non-ideal behavior leading to degradation in the cell efficiency [13]. The ideality factor value depends on irradiance, temperature, and the type of recombination of charge carriers present in valence band and conduction band of the p-n junction diode [14].

2.1.1   ELECTRICAL PARAMETERS OF A PHOTOVOLTAIC CELL/MODULE

An example of the current-voltage (I-V) curve of a photovoltaic module is shown in Figure 2.7. The curve is obtained by using the five-parameter model, discussed in Section 2.2, for a Sharp NT-175U1 photovoltaic module at an irradiance of 900 W/m^2 and module temperature of 50 °C. The Sharp Nt-175U1 is used for all the simulations in this book. Unless otherwise mentioned, the irradiance and module temperature correspond to 900 W/m^2 and temperature of 50 °C. The parameters that determine the photovoltaic module's I-V characteristics are the following.

1. Short-circuit current (I_{SC}): The largest current that a photovoltaic cell can generate is known as the short-circuit current I_{SC}. It is the current generated by the photovoltaic cell when its voltage is zero (Figure 2.7). Ideally, when there are no resistive losses, the current generated by the solar cell is equal to the short-circuit current. For example, the short-circuit current I_{SC} obtained from Figure 2.7 is 4.784 *A*.

2. Open-circuit voltage (V_{OC}): The maximum voltage that can be generated across a photovoltaic cell is known as the open-circuit voltage V_{OC}. It corresponds to the condition when the net current through the photovoltaic cell is zero (Figure 2.7). Substituting $I = 0$ in Equation (2.3) gives

$$V_{OC} = \frac{nkT}{q} \ln\left(\frac{I_L}{I_0} + 1\right).$$

(2.4)

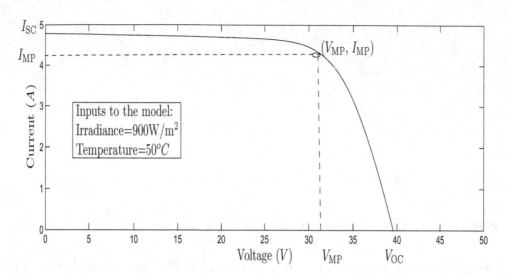

Figure 2.7: I-V characteristic of a photovoltaic module.

This equation shows that V_{OC} varies logarithmically with irradiance. Both light-generated current I_L and dark saturation current I_0 depend on the structure of the device, but I_0 can vary by many orders of magnitude depending on the device geometry and processing [7]. Hence, it is the value of I_0 that determines the open-circuit voltage in practical devices. It can be observed from Figure 2.7 that power is generated only when the voltage is in between 0 and V_{OC}. For voltages outside this range, the device consumes power, instead of supplying it. The open-circuit voltage V_{OC} is obtained as 39.56 V from Figure 2.7.

3. Maximum power point power (P_{MP}): The maximum power P_{MP} produced by the solar cell is reached at a point on the I-V characteristic where the product of voltage and current, IV, is maximum [7]. This current and voltage are known as the maximum power current I_{MP} and maximum power voltage V_{MP}, respectively. Therefore,

$$P_{MP} = V_{MP} I_{MP}. \tag{2.5}$$

From Figure 2.7, V_{MP}, I_{MP}, and P_{MP} are obtained as 31.43 V, 4.33 A, and 136.09 W, respectively.

4. Efficiency (η): The Efficiency (η) of a solar cell is defined as the ratio of the output energy of the solar cell to the input energy from the sun [11]. It is the fraction of incident solar power that can be converted into electricity.

$$\eta = \frac{V_{OC} I_{SC} F}{P_{\text{incident}}}, \tag{2.6}$$

where P_{incident} is the incident solar power and F is the fill factor of a solar cell. The fill factor is defined as the ratio of the maximum power generated by the solar cell to the product of I_{SC} and V_{OC} [11].

$$F = \frac{V_{\text{MP}} I_{\text{MP}}}{V_{\text{OC}} I_{\text{SC}}}. \tag{2.7}$$

Substituting for the fill factor in Equation (2.6),

$$\eta = \frac{V_{\text{MP}} I_{\text{MP}}}{P_{\text{incident}}} = \frac{P_{\text{MP}}}{P_{\text{incident}}}. \tag{2.8}$$

Therefore, efficiency is the ratio of the maximum power generated by the solar cell to the power incident on the solar cell.

2.1.2   RESISTIVE LOSSES IN A SOLAR CELL

Characteristic resistance (R_{ch}) of a solar cell is defined as the resistance of the solar cell at its maximum power point. If the resistance of the load connected to the solar cell is equal to the characteristic resistance, then the solar cell operates at its maximum power point and delivers maximum power to the load. The characteristic resistance is given as

$$R_{\text{ch}} = \frac{V_{\text{MP}}}{I_{\text{MP}}}. \tag{2.9}$$

It is also approximated as [11]

$$R_{\text{ch}} \approx \frac{V_{\text{OC}}}{I_{\text{SC}}}. \tag{2.10}$$

The characteristic resistance is shown in Figure 2.8.

Some of the power generated by the solar cell is dissipated through parasitic resistances. These resistive effects are electrically equivalent to resistance in series R_{s} and resistance in parallel R_{sh} as shown in the Figure 2.9 [11]. The key impact of parasitic resistances is to reduce the fill factor. In the presence of these resistances, the current generated by the solar cell is given by [11]

$$I_{\text{cell}} = I_{\text{L}} - I_0 \exp\left[\frac{q\,(V_{\text{cell}} + I_{\text{cell}} R_{\text{s}})}{nkT}\right] - \frac{V_{\text{cell}} + I_{\text{cell}} R_{\text{s}}}{R_{\text{sh}}}. \tag{2.11}$$

The series resistance R_{s} is the resistance offered by the material of the solar cell to the current flow [10]. Its main effect is to reduce the fill factor and therefore the efficiency of cell. Excessively high values of R_{s} may also reduce the short-circuit current [11]. On the other hand, the shunt resistance R_{sh} is a result of the manufacturing defects in the solar cell. Low shunt resistances provide an alternate current path for the light-generated current, thereby reducing the terminal voltage of the solar cell. This effect is more pronounced at lower irradiances, since there will be less light generated current [11]. For an efficient solar cell, R_{s} should be as small and R_{sh} as large as possible.

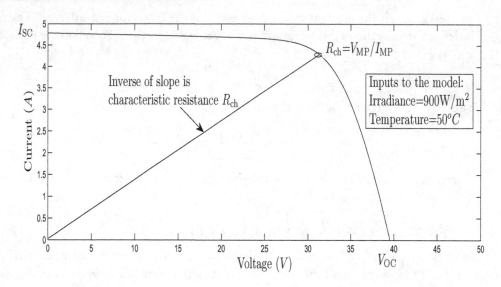

Figure 2.8: I-V curve obtained from five-parameter model at an irradiance of 900 W/m^2 and temperature of 50°C.

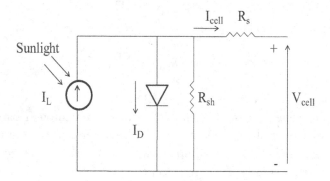

Figure 2.9: Parasitic series and shunt resistances in the equivalent circuit of a solar cell.

2.1.3 BYPASS DIODES IN A PV MODULE

Bypass diodes are used to avoid "hot-spot" conditions in PV arrays [11]. A hot-spot occurs when one or more PV modules (or cells) are shaded (or partially shaded) in a string. These shaded modules will be forced by the remainder of the array to operate above their short-circuit current, consuming rather than generating power. This heats up the module and forms a hot-spot. Connecting bypass diodes across groups of solar cells in a module prevents the occurrence of hot-spots by limiting the power dissipated by a shaded module. The operation of bypass diodes and their effect on the electrical behavior of shaded PV modules is explained in Chapter 3.

2.1.4   EFFECT OF TEMPERATURE AND IRRADIANCE

The power output of solar cells is significantly affected by variations in the temperature and irradiance. This section describes their impact on the characteristics of the solar cell.

Temperature

The most significant effect of temperature is on the cell terminal voltage V_{cell}. It decreases with increase in temperature, i.e., it has a negative temperature coefficient. The impact of temperature on current is less pronounced. Figure 2.10 shows the effect of temperature on the I-V characteristic at a constant irradiance of 1000 W/m^2.

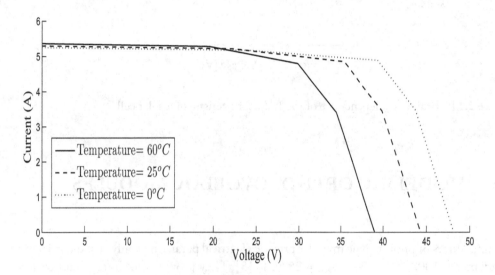

Figure 2.10: Temperature dependence of the I-V characteristic of a solar cell.

Irradiance

The I-V characteristics of solar cell under different levels of illumination are shown in Figure 2.11. It is observed that the light generated current I_L is directly proportional to the irradiance. Therefore, the short-circuit current is directly proportional to the irradiance. The voltage variation is much smaller due to its logarithmic dependence on the irradiance and is often neglected in practical applications [8]. Figure 2.11 shows the effect of irradiance on the I-V characteristic at a constant temperature of 25°C.

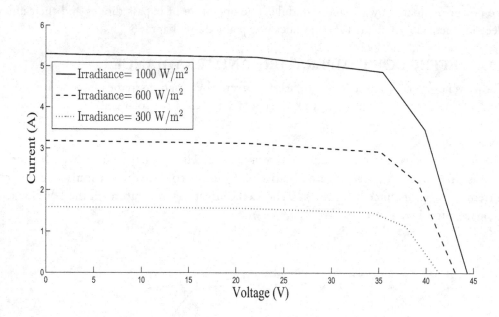

Figure 2.11: Irradiance dependence of the I-V characteristic of a solar cell.

2.2   MODELING OF PHOTOVOLTAIC MODULES

Manufacturers of photovoltaic modules provide electrical parameters only at standard test conditions (irradiance = 1000 W/m^2 and T_{cell} = 25 °C) [15]. They provide the short-circuit current I_{SC}, the open-circuit voltage V_{OC}, the voltage at maximum power point V_{MP}, the current at maximum power point I_{MP} and the temperature coefficients at open-circuit voltage and short-circuit current [15]. The nominal operating cell temperature determined at an irradiance of 800 W/m^2 and an ambient temperature of 20°C is also specified [6]. However, PV modules operate over a large range of conditions and the information provided by manufacturers is not sufficient to determine their overall performance. This motivates the need for an accurate tool to determine the module behavior. Photovoltaic performance models are therefore built to predict the performance of a module at any operating condition. A PV performance model predicts the I-V characteristic of a PV module as a function of temperature, incoming solar irradiation (direct and diffuse), angle of incidence, and the spectrum of sunlight. Angle of incidence (AOI) is the angle between a line perpendicular (normal) to the module surface and the beam component of sunlight [3]. Models are used to compare an array's predicted and measured performance and detect problems that affect the modules' efficiency [3].

The most common metric comparing measured module performance to predicted behavior is the performance ratio [16]. This figure is the ratio of the PV array's final system yield Y_f to the reference yield Y_r:

$$PR = Y_f / Y_r. \tag{2.12}$$

Y_f is defined as the ratio of energy produced over the time period of interest E to the rated power of the array P_0: $Y_f = E/P_0$. Y_f has units of hours but is more typically written in terms of kWh/kW. It represents the amount of time an array would have to operate at its rated power in order to produce the same energy. Y_r, on the other hand, compares the available solar energy per unit area H over the same time period to the reference irradiance G_{ref}, typically 1000 W/m^2. That is, $Y_r = H/G_{ref}$. Y_r also has units of hours and expresses the available solar energy as an equivalent number of hours at the reference irradiance. The performance ratio, then, compares the energy produced by the array to the energy produced by a hypothetical ideal array whose output varies linearly with irradiance and depends on no other factors. It is the standard for evaluation of PV arrays when detailed performance models are not available.

In this section models that determine the module's behavior on both the DC and AC side of the inverter are discussed. The Sandia model and five-parameter model described here are widely used to predict the module's performance on the DC side of the inverter.

2.2.1   THE SANDIA MODEL

Sandia National Laboratories has developed a photovoltaic module and array performance model [3]. It uses a database of empirically derived parameters developed by testing modules from a variety of manufacturers to predict photovoltaic module/array performance [17].

The Sandia model is based on a set of equations that describe the electrical performance of the photovoltaic modules. These equations can be used for any series or parallel combination of modules in an array. They calculate the four points necessary to define the I-V curve of the photovoltaic module/array. These are the short-circuit current (I_{SC}), the open-circuit voltage (V_{OC}), the voltage at maximum power point (V_{MP}), and the current at maximum power point (I_{MP}). Two other currents are calculated at intermediate values for modeling the curve shape. These are defined at a voltage equal to half of the open-circuit voltage and at a voltage midway between the voltage at maximum power point and open-circuit voltage. All these parameters are found by a curve fitting process of the coefficients obtained from testing of the modules. Empirical coefficients are also developed to determine parameters that are temperature dependent, effects of air mass and angle of incidence on the short-circuit current and type of mounting (whether rack-mounted or building-integrated PV systems) [17]. This model also determines the effective irradiance, defined as the fraction of the total irradiance incident on the modules to which the cells actually respond (dimensionless or "suns") [3]. Effective irradiance is used in the calculation of model's parameters.

An I-V curve generated by the Sandia model and programmed in MATLAB is shown in Figure 2.12. The program for Sandia model is provided in the Appendix.

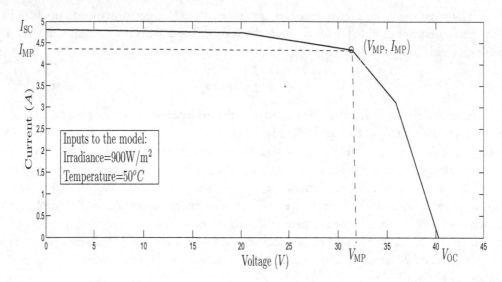

Figure 2.12: I-V curve obtained from Sandia model at an irradiance of 900 W/m^2 and temperature of 50°C.

The Sandia model was validated using experimental data from different geographic locations provided by the National Institute for Standards and Technology (NIST) [17]. The model can be used to predict output power within 1% of the measured power and is considered to be the most accurate of all the available performance models [17]. However, it has limitations in that it requires several parameters that are not available from the manufacturer which necessitates additional testing of the modules.

2.2.2   THE FIVE-PARAMETER MODEL

The five-parameter model developed at the University of Wisconsin Solar Energy laboratory uses the well-known single diode model shown in the Figure 2.9 to predict the performance of PV modules. The five-parameter model predicts the maximum power and I-V characteristics of a photovoltaic module at any operating condition. Unlike the Sandia model, it utilizes only data provided by the manufacturer at standard test conditions and does not require additional measurements do derive parameters. This model has five parameters: the light-generated current I_L, the diode reverse saturation current I_0, the series resistance R_s, the shunt resistance R_{sh} and the modified ideality factor a (see Figure 2.9) [15]. The modified ideality factor in terms of ideality factor n is given as

$$a = \frac{N_s n k T_{cell}}{q}, \tag{2.13}$$

where N_s represents the number of cells in series.

From Figure 2.9, the current generated by the photovoltaic module (I_{mod}) is given as [6],

$$I_{mod} = I_L - I_0 \left[\exp \left(\frac{V_{mod} + I_{mod} R_s}{a} \right) - 1 \right] - \frac{V_{mod} + I_{mod} R_s}{R_{sh}}, \qquad (2.14)$$

where V_{mod} is the terminal voltage of the module. The five-parameter model is semi-empirical as it calculates parameters theoretically, from known relationships and equations derived from previous studies [17]. The model first calculates the reference parameters, i.e., the parameters at standard test conditions, using the available manufacturer's data. These values are used to find the parameters at any operating condition. An easy to use application to determine the reference parameters was developed by the University of Wisconsin [18]. A sample I-V curve of the five-parameter model implemented in MATLAB is shown in Figure 2.7.

The five-parameter model is compared with Sandia model and experimental data provided by National Institute of Standards and Technology (NIST) in the following section.

2.2.3   COMPARISON OF THE SANDIA AND THE FIVE-PARAMETER MODELS

The experimental data provided by the National Institute of Standards and Technology (NIST) is used to examine the efficiency of the five-parameter and Sandia models [15]. It is observed that the Sandia model shows slightly better agreement with the data in contrast with the five-parameter model [15]. This is expected as the Sandia model requires many measurements over a wide range of conditions to determine the model parameters (as seen in section 2.2.2), whereas the five-parameter model uses only the manufacturer's data to find the I-V characteristics. The five-parameter model can give better results if I-V curves corresponding to two different reference conditions are provided instead of just one at STC (for example one set at STC and the other at 500 W/m^2 irradiance and 40°C module temperature) [15]. It is also to be noted that there are uncertainties inherent in the experimental data.

2.3   PHOTOVOLTAIC ARRAY TOPOLOGIES

Photovoltaic cells are electrically connected to form a Photovoltaic module. The schematic symbol for a photovoltaic cell or module is shown in Figure 2.13. Photovoltaic modules are interconnected in series-parallel combinations to form a photovoltaic array as shown in Figure 2.14. The limits on the number of modules to connect in series and parallel is discussed in Chapter 5. This section discusses the electrical characteristics of a photovoltaic array under ideal conditions.

In Figure 2.14 the photovoltaic array consists of L modules connected in series to form a string and N such strings are connected in parallel. Each photovoltaic module is represented by $M_{i,j}$, where i and j represent the row and column index, respectively. The current and voltage of a module $M_{i,j}$ are represented by $I_{i,j}$ and $V_{i,j}$, respectively. The modules present in a string carry same amount of current and the string current equals the module current. The string voltage is the sum of the voltages of individual modules in the string. The string current and voltage are obtained

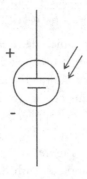

Figure 2.13: Photovoltaic module symbol.

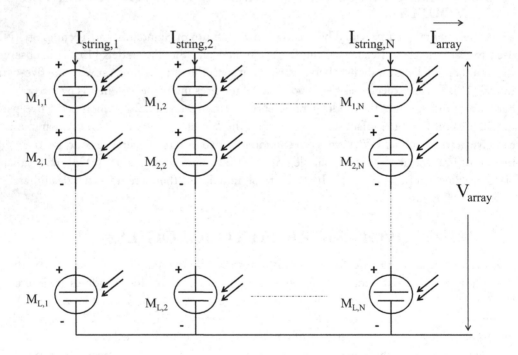

Figure 2.14: Photovoltaic modules connected in series-parallel (SP) configuration.

as

$$I_{\text{str},j} = I_{i,j} = I_{k,j} \, , \tag{2.15}$$

$$V_{\text{str},j} = \sum_{i=1}^{L} V_{i,j} \, , \tag{2.16}$$

where $j = 1, 2, ..., N$. When strings are connected in parallel to form an array, the array current equals the sum of the currents from each string. The array voltage is same as the voltage of any string voltage. The current and voltage of an array are obtained as

$$I_{\text{arr}} = \sum_{j=1}^{N} I_{\text{str},j} \tag{2.17}$$

$$V_{\text{arr}} = V_{\text{str},j} = V_{\text{str},k}. \tag{2.18}$$

The power of the array is obtained as

$$
\begin{aligned}
P_{\text{arr}} &= V_{\text{arr}} I_{\text{arr}} \\
&= (L V_{1,1})(N I_{1,1}) \\
&= L N P_{1,1} \, ,
\end{aligned}
$$

where $P_{1,1}$ represents power of the module $M_{1,1}$. Therefore, under ideal conditions, the array power output is equal to the sum of the powers of individual modules.

Apart from the series-parallel (S-P) configuration, the modules can also be connected in a cross-tied manner in which additional connections are introduced between the modules. There are two kinds of cross-tied topologies: the total cross-tied (TCT) topology and the bridge link (BL) topology. In the total cross-tied topology as shown in Figure 2.15, each of the photovoltaic modules is connected in series and parallel with the others [19]. The bridge link topology shown in Figure 2.16, consists of half of the interconnections when compared to the total cross-tied topology [19]. Ideally, when there are no wiring losses and module mismatches, all the modules behave identically and the performance (the generated array power) is the same for the series-parallel and cross-tied topologies. When there are electrical mismatches, one of the topologies outperforms the others. The performance of photovoltaic arrays in the presence of electrical mismatches and faults is discussed in Chapter 3. The S-P configuration is nearly always used in practice, as it requires fewer connections and uses less cable.

2.4   SUMMARY

This chapter gives a general view of the operation of photovoltaic cells and their electrical behavior. The electrical equivalent circuit of the photovoltaic cell is presented and the electrical parameters of the I-V characteristic are explained. The effect of temperature and irradiance on the performance

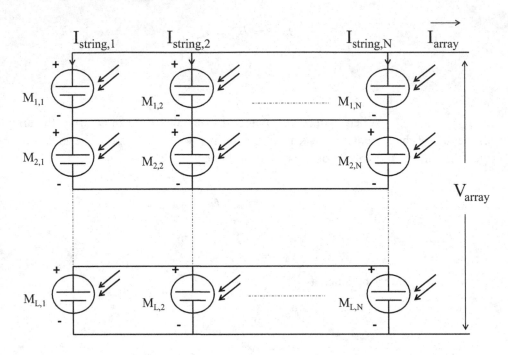

Figure 2.15: Photovoltaic modules connected in total cross-tied (TCT) configuration.

of a solar cell is discussed. The need for a photovoltaic model and the existing accurate models that predict the performance of photovoltaic systems are explained. The array topologies used in practice are presented and their electrical characteristics under ideal conditions are explained.

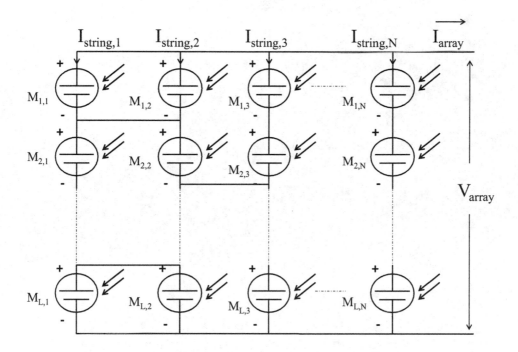

Figure 2.16: Photovoltaic modules connected in bridge Link (BL) configuration.

CHAPTER 3

Causes Performance Degradation and Outage

In an ideal solar array, array power is simply the sum of individual module powers. However, several conditions result in available DC power from the array being significantly below predicted levels. Some of these effects, such as mismatches, occur in all arrays at all times. Others are intermittent faults which are repaired as soon as they are detected. This chapter examines the effects of some of the most common non-ideal array behaviors.

3.1 SHADING

Shading, the total or partial blockage of sunlight from a PV module surface, is a very serious concern in PV arrays [20, 21, 22]. It causes large performance drops and can even damage modules if not properly controlled. When a PV cell is shaded, its light-generated current I_L (see Figure 2.9) decreases, causing short-circuit current I_{SC} and maximum power current I_{MP} to drop dramatically. In the extreme case of 100% shading, $I_{SC} = I_{MP} = 0$ A, and the PV cell acts as a diode. When a shaded cell is connected to unshaded cells in a series-parallel configuration, the shaded cell's maximum current I_{SC} is significantly less than the optimal current I_{MP} of the unshaded cells. Since each cell in the string must conduct the same current, the entire string is constrained to operate at the short-circuit current of the shaded cell, severely restricting the current and therefore power available from the remaining unshaded cells. A similar effect occurs at the module level. This effect is seen in Figure 3.2(a), which shows the simulated I-V characteristics of a series string of two Sharp NT-175U1 PV modules at standard test conditions (STC) without bypass diodes, where one module has been shaded reducing its incoming irradiance to 200 W/m^2.

3.1.1 BYPASS DIODES

In order to mitigate the effects of shading and other mismatches, bypass diodes are used (Figure 3.1). The bypass diodes D_{B1} and D_{B2} are connected such that under normal conditions ($V_{cell} \geq 0$), they are reverse biased by the PV cell and conduct no current. However, when the PV cell itself is reverse biased by some external source ($V_{cell} < 0$), the bypass diode activates, conducting large currents at low voltages. This limits the reverse voltage, and therefore power dissipation, of the shaded cell.

The heavy line in Figure 3.1 shows the dominant current flow in the case of severe shading of one cell in a two-cell string. Shading causes I_{L2} to drop to a very low value while I_{L1} remains large.

D_{B2}, the bypass diode of the shaded cell, is forward biased and conducts the additional current that is no longer able to flow through I_{L2}. Of the remaining diodes, D_{B1} and D_{C2} are reverse biased, and D_{C1} is forward biased but operates below its threshold voltage, and therefore conducts relatively small currents. In practice, a single bypass diode is usually connected across multiple cells, with similar effect.

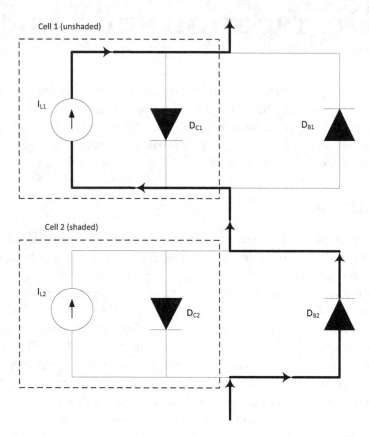

Figure 3.1: Schematic of bypass diode operation under shading conditions.

Figure 3.2(b) shows the I-V characteristic of a series string of two Sharp NT-175U1 PV modules with a bypass diode for each module at standard test conditions (STC), one of which has been shaded by reducing its incoming irradiance to 200 W/m². In contrast to Figure 3.2(a), the unshaded module is allowed to operate at or near its maximum power point (MPP). This reverse biases the shaded cell, activating its bypass diode. The effect of the bypass diode may be seen in the I-V curves of the individual modules: while the current of a module with no bypass diode is effectively limited to I_{SC} even at large reverse voltages, the addition of a bypass diode causes the combined I-V curve of the module and diode to bend sharply upward in the reverse voltage ($V_{cell} < 0$) region.

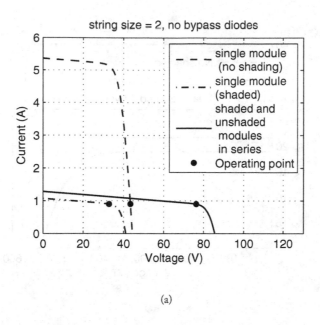

(a)

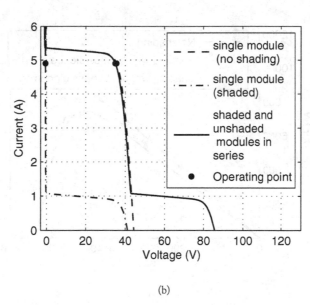

(b)

Figure 3.2: Effect of shaded (1/5 irradiance) module on IV curves for two-module string (a) without and (b) with bypass diodes.

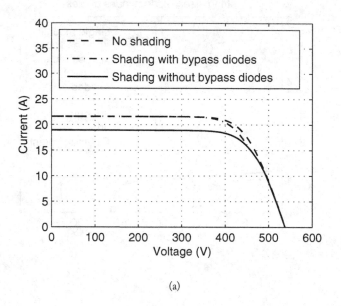

(a)

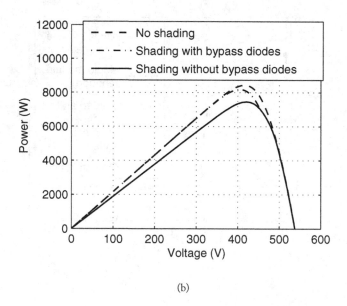

(b)

Figure 3.3: Effect of a single shaded (1/2 normal irradiance) module on (a) I-V curve and (b) P-V curve.

The action of a bypass diode can be easily understood for a single string of two modules, but what about a full array? Figure 3.3 simulates the effect of a single shaded (500 W/m^2 irradiance) module in a 13 × 4 array of Sharp NT-175U1 modules at STC. Note that when bypass diodes are present, the effect of shading on output power is dramatically reduced. However, when bypass diodes are not present, current in the string where shading appears is severely limited, creating a disproportionately large drop in power relative to the shading effect. In addition to this problem, the shaded module is severely reverse biased ($V = -27$ V) at the MPP while still conducting its maximum (short-circuit) current of 2.7 A. This large reverse bias and current causes the module to dissipate power, creating a hot spot which may cause damage to the array. It is because of these effects that virtually all modern PV arrays are equipped with bypass diodes. Figure 3.3 shows another important point: partial shading causes a disproportionately large drop in array output power relative to the decrease in irradiance. Even with bypass diodes, a 1% reduction in available input power resulted in a 3% reduction in output power due to partial shading. Without bypass diodes, this effect is even worse, causing a 12% drop in output power. For this reason, great care is taken in the design of PV arrays to minimize the time spent under shading conditions.

3.2   MODULE MISMATCH

Every type of PV module has variable characteristics inevitably caused by process variation; the optimal current and voltage will not be the same for each module in an array at a given point in time. These variations have the effect of reducing the output of the array, since the current and voltage of a module are constrained by the array's electrical configuration. Module mismatch causes each module to operate at a suboptimal point on the current-voltage (I-V) curve, reducing the array's power output. While improvements in manufacturing processes have reduced the effect of these variations, they still must be considered [23]. This section describes the effects of variation in the parameters of the single-diode model on the series-parallel (S-P) array configuration, in which modules are arranged into series "strings" that are then connected in parallel.

3.2.1   MISMATCHES IN CURRENT

One simple source of mismatch in a PV array is uncertainty in the light generated current I_L of each module, defined in Figure 2.9; this mismatch is especially harmful in the series-parallel configuration. The effects of additive zero-mean Gaussian variation in I_L are simulated for a 13-series, 4-parallel (13 × 4) array of Sharp NT-175U1 solar modules in Figure 3.4. Gaussian variation was selected because variation in I_L is expected to be a sum of many relatively independent factors; the authors are not aware of any work which attempts to determine the distribution of modules' I_L values. The light current of the i-th module $I_{L,i}$ was determined as follows:

$$I_{L,i} = I_L + w(i) \tag{3.1}$$

where $w(i) \sim \mathcal{N}(0, \sigma^2)$ is a Gaussian random variable with mean 0 and variance σ^2, for the cases of $\sigma = 0.2I_L$ and $\sigma = 0.05I_L$. The 20% tolerance ($\sigma = 0.2I_L$) case has been selected in order to highlight the effect of mismatches; modern PV modules have a (not necessarily Gaussian) tolerance of I_L on the order of 3–10%, causing a power loss of up to 4% relative to a perfectly matched array [23].

It can be seen that in every case the presence of mismatches reduces the power output of the array. This is due to the inherent limitations of the electrical configuration. Each module in a string must carry the same current, forcing the individual modules to operate at a slightly sub-optimal point on their I-V curve. A similar but less important effect occurs between strings, which must each develop the same voltage. An interesting effect is visible in Figure 3.4 for the case of 20% standard deviation in the S-P configuration: the I-V curve develops corners, in turn causing local maxima in the power-voltage (P-V) curve. These corners appear when one or more modules becomes reverse biased, activating its bypass diodes and effectively turning it into a consumer of energy. At the maximum power point (MPP), the worst performing modules are effectively off-line due to mismatch. The effects of a zero-mean additive current mismatch are, on average, always harmful to the power output of the array. This is demonstrated in Figure 3.4(b), which shows the averaged PV curve of 10 independent simulations of the $\sigma = 0.2I_L$ case outlined above in (3.1). Note that while the presence of zero-mean variation in I_L does not affect the expected maximum power output of a single module, it lowers the expected maximum power of the array, in this case from 8.4 kW to 6.5 kW.

3.2.2   MISMATCHES IN DIODE PARAMETERS

Uncertainty in diode parameters of saturation current I_0 and ideality factor n, defined in Section 2.1, leads to variation in the voltage of modules. Figure 3.5 shows the effect of a 20% tolerance in n and I_0 for the case of a 4-module array with all 4 modules in parallel. As in Section 3.2.1, variation in parameters was modeled as additive Gaussian noise. This parallel configuration maximizes the effect of mismatch, since with a string size of 1, high-voltage modules are not able to compensate for low-voltage modules in the same string. In the S-P configuration, however, this variation has only a minimal effect on array power output. The voltage of a string is the sum of the voltages of its modules, and even if the maximum power voltage V_{MP} of individual modules varies significantly, the V_{MP} of a 13-module string will still have negligible variation relative to the total voltage of the string. Because of this, mismatches in module V_{MP} are much less important than mismatches in I_{MP}.

3.3   MODULE SOILING

Module soiling is the build-up of dirt on the surface of a PV module. Researchers have found that the effects of soiling are relatively small (2.3% loss of power) for directly incident light but become more significant for larger angles: an 8.1% loss was observed in a soiled module when light is incident from an angle of 56° [24]. It was also found that bird droppings cause a much larger degrading effect than general soiling due to their complete blockage of light over a small area and can be treated as

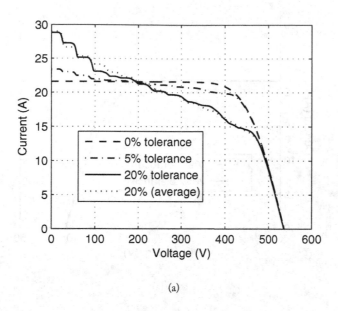

(a)

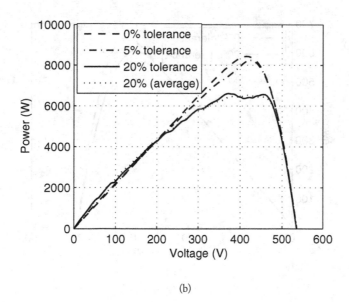

(b)

Figure 3.4: Effect of mismatch in I_L on (a) I-V curve and (b) P-V curve.

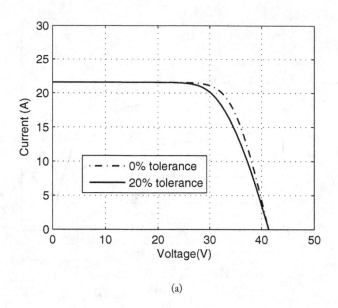

(a)

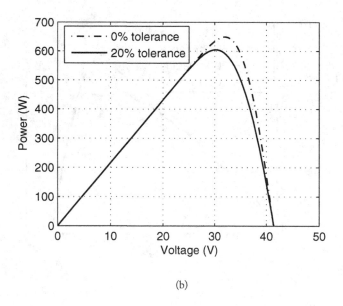

(b)

Figure 3.5: Effect of mismatch in diode parameter *n* on (a) I-V curve and (b) P-V curve.

a partial shadow (see Section 3.1). If one module is soiled, it will cause a mismatch in the effective irradiance of the modules, reducing the array performance.

3.4   GROUND FAULT

A ground fault occurs when the circuit develops an unintentional path to ground. This results in lowered output voltage and power, and can be fatal if the leakage currents are running through a person. AC Ground fault circuit interrupters (GFCIs) are capable of detecting even very small (6 mA) leakage currents and opening a switch to stop current flow within 200 ms. Actual detection thresholds and response times vary widely by application, and there is generally a trade-off between response time and detection threshold. Typical AC GFCIs work on the principle of detecting mismatches between incoming and outgoing current: all wires going to and from a voltage source pass through the sensing opening of a current transducer. Under normal operation, all currents must pass through the transducer, meaning that the net current through the transducer is zero. Under fault conditions, some current does not pass through the detector, instead flowing through the faulty path and creating a non-zero net current through the transducer's opening. When this current is detected, a relay is quickly opened and all current flow is stopped [25] (Figure 3.7). GFCIs are a mature technology.

In a PV array, however, the existence of multiple power sources and potential fault locations means that a traditional GFCI which simply detects net current flow will be unable to detect many potential fault scenarios. Instead, ground fault detection in a PV array typically takes the form of a simple over-current detector with a detection threshold of 0.5–1 A [26]. When a fault occurs, extra current in the grounding system is detected and the circuit is opened. GFCIs are mandated by the 2008 United States National Electric Code for all PV systems [27]. Figure 3.6 shows the effect of a severe ground fault on the currents and voltages of modules in a series-parallel configuration.

3.5   DC ARC FAULT

Direct Current (DC) arcing is a spark across air or another dielectric and occurs in two forms: series and parallel arcs [28, 29, 30]. A series arc most often occurs when a connection breaks, leaving two conductors very near each other. Series arcs can occur in junction boxes, at the cable connections between modules, and within modules. Parallel arcs can occur when two conductors of different voltage are near each other, such as when the insulation of two wires running parallel to one another is compromised. The current produced by a series arc is limited by the load with which it is in series, while a parallel arc can consume as much current as a source is able to supply. Arcs can lead to inefficiency in array operation, frequently cause failure of bypass diodes [31], and can even cause fires. Even a relatively benign arcing incident is likely to disable the entire string in which the arc occurs.

Unlike other faults mentioned in this chapter, arc faults are an inherently transient phenomenon and cannot be fully modeled by a change in the DC behavior of an array. The DC

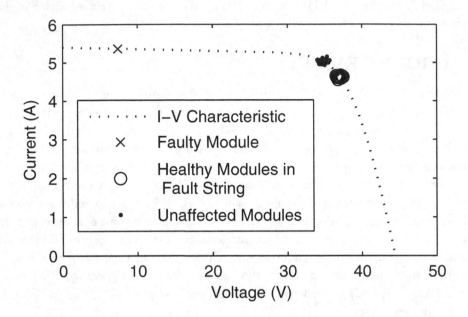

Figure 3.6: Module voltage and current during ground fault conditions.

behavior of arcs can be modeled as a nonlinear resistance [32], but more complex techniques are needed for transient behavior. Two intuitive and long-standing arc models are the Cassie [33] and Mayr [34] models, which describe an arc as a time-varying resistance. The Cassie model assumes that the arc is a cylinder of conductive plasma known as the arc column. Constant current density in the arc column, convective heat loss through its entire volume, and constant resistivity and energy storage density in the arc are also assumed. These assumptions result in an arc equation of the form

$$\frac{1}{R_{\text{arc}}(t)} \frac{d R_{\text{arc}}(t)}{dt} = \frac{1}{\tau} \left(1 - \frac{V_{\text{arc}}(t)^2}{V_0^2} \right) , \tag{3.2}$$

where $R_{\text{arc}}(t)$ is the instantaneous arc resistance, τ is the arc time constant, $V_{\text{arc}}(t)$ is the voltage across the arc, and V_0 is the steady-state arc voltage. Equation 3.2 may equivalently be written in terms of the arc conductance $G_{\text{arc}}(t)$:

$$G_{\text{arc}}(t) = \frac{V_{\text{arc}}(t) I_{\text{arc}}(t)}{V_0^2} - \tau \frac{d G_{\text{arc}}(t)}{dt} , \tag{3.3}$$

where $I_{\text{arc}}(t)$ is the current through the arc and other quantities are defined as before. At steady state, the Cassie model implies a constant arc voltage regardless of current. While valid for large currents, the model breaks down at low currents and does not allow for ignition or extinction of an arc.

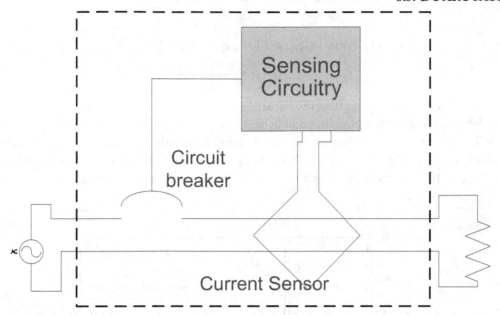

Figure 3.7: Simplified block diagram of GFCI operation.

The Mayr model, on the other hand, is accurate at lower currents and accounts for arc extinction. It assumes constant power dissipation, P_0, from the edge of the arc column and assumes the conductance of the arc to be proportional to the energy stored in the arc column. The differential equation defining the Mayr arc model is

$$\frac{1}{R_{\text{arc}}(t)} \frac{d R_{\text{arc}}(t)}{dt} = \frac{1}{\tau} \left(1 - \frac{V_{\text{arc}}(t) I_{\text{arc}}(t)}{P_0} \right) \tag{3.4}$$

and may be equivalently re-written in terms of the conductance as

$$G_{\text{arc}}(t) = \frac{I_{\text{arc}}(t)^2}{P_0} - \tau \frac{d G_{\text{arc}}(t)}{dt}. \tag{3.5}$$

At steady state, the Mayr model's I-V characteristic is a hyperbola. The model is not accurate at large currents where convective heat dissipation from within an arc becomes a significant effect.

Many related hybrid Cassie-Mayr arc models have been proposed [35] in an attempt to more completely describe the arc characteristics. In [32], a weighted sum of the Cassie and Mayr arc conductances is taken. The weights vary as a function of the arc current, with a very small minimum conductance $G_{\min}$ added to allow for self-ignition of the arc. The conductance of this hybrid model is given by:

$$G_{\mathrm{arc}}(t) = \quad G_{\min} + \left[1 - \exp\left(\frac{-I_{\mathrm{arc}}(t)^2}{I_t^2}\right)\right] \frac{V_{\mathrm{arc}}(t) I_{\mathrm{arc}}(t)}{V_0^2} \tag{3.6}$$
$$+ \exp\left(\frac{-I_{\mathrm{arc}}(t)^2}{I_t^2}\right) \frac{I_{\mathrm{arc}}(t)^2}{P_0} - \tau \frac{dG_{\mathrm{arc}}(t)}{dt},$$

where I_0 is a constant representing the transition point between the two models and other parameters are defined as before. Another common hybrid model is the Habedank model [36], which defines a Cassie arc in series with a Mayr arc. Figure 3.8 shows the steady-state current-voltage characteristics of the Cassie and Mayr models given above, along with the Habedank model. Parameters were chosen as $G_{\min} = 10^{-8}$ S, $P_0 = 30$W, and $V_0 = 28$ V. The parameter τ does not affect the steady-state I-V characteristic.

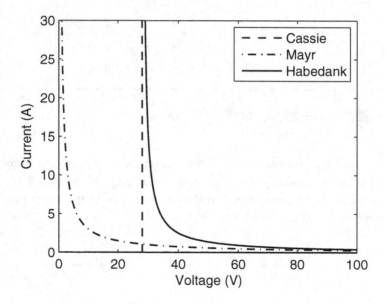

Figure 3.8: Comparison of Cassie, Mayr, and hybrid models.

3.6 HIGH-RESISTANCE CONNECTIONS

A high-resistance connection is a small region of high resistance through which current must flow, causing unintended heating and power loss. High-resistance connections may be formed within a module during the manufacturing process, during array assembly if connections are not made tightly, or over time due to factors such as thermal stresses [37]. A high-resistance connection may eventually separate fully, leading to a series arc fault or open circuit. High-resistance connections increase the effective series resistance of a module, leading to reduced power output. In extreme cases, a bad connection may heat up enough to be hazardous, leading to the colloquial name "glow fault" to refer to such a connection.

3.7   INVERTER FAILURE

Inverters are generally the weakest link in a PV array, and inverter failures are the most common cause of PV array outages. The failure of switching transistors is the most common cause of inverter failure, although degradation of electrolytic capacitors also plays a large role [38]. Several approaches are taken to mitigate the risk of inverter failure. These include designs which limit the use of electrolytic capacitors, selection of over-rated switching transistors, and the use of multiple, smaller inverters in place of a single large one. Thermal stresses are the most significant factor leading to inverter failure, so the design of cooling systems is extremely important for inverter reliability. Many modern inverters report their operating status, including efficiency, to a server for logging and monitoring. This information better allows an array operator to quickly identify and repair faulty inverters.

3.8   ISLANDING

The addition of arrays to the power grid poses a safety risk known as islanding. Islanding occurs when a fault in the wider electrical grid causes a power outage, but a PV array does not turn off. In this situation, a small "island" forms in the grid, receiving electrical power from the PV array. This island of energized wires poses a threat to repair technicians seeking to fix the fault. Modern inverters are, for the most part, equipped with anti-islanding features which detect when the AC grid goes down and shut off the inverter. In addition to this safeguard, many electrical companies require that residential arrays connected to their grid come equipped with a manual disconnect switch, to be opened by a power company technician before beginning work on nearby parts of the grid.

3.9   SUMMARY

The preceding sections have given an indication of the types of faults a PV array experiences and the effect they have on array IV curves. Throughout the cases examined it was found that faults occurring in individual modules have far-reaching effects on array behavior: PV modules are not independent units whose performance can be easily compared to one another. It is this interdependence of module output in the series-parallel configuration which makes the problem of detecting, localizing, and diagnosing faults non-trivial.

CHAPTER 4

Fault Detection Methods

In this chapter, several signal processing methods are presented that may be used to identify faults of the types presented in Chapter 3. Because PV faults have highly non-local non-independent effects on the array, some sophistication is required in detection methods. However, literature on automated fault detection in PV arrays is relatively sparse. First, several methods from the literature are presented and discussed. Next, classical distance-based methods of outlier detection are considered. Finally, more modern machine learning-based methods are described.

4.1 CURRENT METHODS IN PHOTOVOLTAICS

There are several articles in the literature on early fault detection in solar arrays [39, 40, 41, 42]. In general, two approaches are presented. In the first approach, a photovoltaic array performance model is used to derive the expected I-V curve of an array. This expected I-V curve is compared with the measured operating point of the array. Stellbogen [43] developed such a system. It is capable of detecting deviations of 10% or more in the array output and is also able to predict the effects of shadowing using 3D geometric models. However, future work on this approach must incorporate feedback from array measurements in order to adjust the performance model for natural and inevitable array degradation. Without this adjustment, any such detector will show worsening performance as an array ages.

The second approach to fault detection involves taking multiple measurements within the array and using statistical methods to detect outliers, which are then labeled as faulty. Zhiqiang and Li [44] described an ad-hoc method wherein the current and voltage of each string are measured. A string is determined to be faulty if its current is less than 80% of the highest measured string current. A fault was also identified if a string's current deviation ratio exceeded 5. Deviation ratio was defined as the ratio of $\Delta I / \Delta I_{min}$, where ΔI is the difference in current between the present measurement and the previous one and ΔI_{min} is the minimum measured ΔI among all strings. This method of fault detection makes no attempt to determine the cause of a fault.

Vergura et al. [39] applied Analysis of Variance (ANOVA) and the Kruskal-Wallis test to the problem of detecting performance outliers in solar arrays based on measurements taken at the inverter. However, their work is specific to the case of multiple sub-arrays, each connected to its own inverter, and can only localize faults to the sub-array level. Since the output of sub-arrays can be reasonably assumed to be independent and identically distributed, these methods perform well, especially in predicting inverter failures. However, within an array, module currents and voltages are highly non-independent and other methods are needed.

4.2   ANOMALY DETECTION AND STATISTICS

Anomaly detection, the practice of identifying subsets of data which do not exhibit the same patterns as the rest of the set, is an active area of research with several successful commercial applications. The most notable commercial successes are probably fraud detection algorithms, which monitor financial transaction data for events such as identity theft, and intrusion detection algorithms, which attempt to identify unauthorized use of computer networks. With the availability of higher-resolution data from PV modules and inverters, an opportunity exists for anomaly detection methods to be adapted and applied to PV technology. This section presents some of the most commonly used anomaly detection algorithms. In the examples below, detection is performed based only on module-level current and voltage measurements, although other data such as temperature measurements could be used as well.

Many of the techniques presented in this chapter rely on some measure of multivariate distance between data points or on a measure of the centroid of the dataset. In many cases, the appropriate choice of distance depends on the specific application being considered. This section presents three common distance metrics, the Euclidean, Mahalanobis, and MCD-based distances, and their associated centroid estimators.

4.2.1   THE EUCLIDEAN DISTANCE

Euclidean distance is the traditional measure of distance from geometry, extended to n-dimensional space. In our case, $n = 2$, with current and voltage comprising the dimensions of the data. The Euclidean distance between two points $\mathbf{x}_1$ and $\mathbf{x}_2$ is defined as

$$d_{EU}(\mathbf{x}_1, \mathbf{x}_2) = \|\mathbf{x}_2 - \mathbf{x}_1\|_2 = \sqrt{(\mathbf{x}_2 - \mathbf{x}_1)^T (\mathbf{x}_2 - \mathbf{x}_1)}. \tag{4.1}$$

Euclidean distance is extremely intuitive and is an appropriate distance metric for a few well-scaled problems where the elements of $\mathbf{x}_i$ are relatively uncorrelated. However, Euclidean distance is not scale-invariant. If one element of $\mathbf{x}_i$ measures current on the order of amps and another measures voltage on the order of hundreds of volts, the Euclidean distance will depend almost entirely on the difference in voltage between two data points, while ignoring potentially significant data from current measurements. Mahalanobis distance, discussed in Section 4.2.2 is one attempt to correct for these effects.

4.2.2   THE MAHALANOBIS DISTANCE

Mahalanobis distance is one of a family of distance metrics which attempt to overcome the problems of the Euclidean distance by replacing the ℓ_2-norm with a quadratic norm. In the case of the Mahalanobis distance, the quadratic norm is defined by the covariance matrix $\mathbf{C}$ of the dataset:

$$d_{MA}(\mathbf{x}_1, \mathbf{x}_2) = \sqrt{(\mathbf{x}_2 - \mathbf{x}_1)^T \mathbf{C}^{-1} (\mathbf{x}_2 - \mathbf{x}_1)}. \tag{4.2}$$

Compared with the Euclidean distance, Mahalanobis distance offers several important advantages. It is scale invariant — in fact, any non-singular linear transformation may be applied to the data set without altering $d_{MA}(\mathbf{x}_1, \mathbf{x}_2)$: by recalculating C, the effects of the transformation will be canceled out. Another effect of this invariance is that correlations between the elements of $\mathbf{x}_i$ are automatically compensated for. Taking the Mahalanobis distance between two points is equivalent to taking the Euclidean distance after the dataset has undergone centering, decorrelation, and normalization procedures.

Mahalanobis distance is commonly used to measure $d_{MA}(\mathbf{x}_i, \boldsymbol{\mu})$, the distance between the sample centroid $\boldsymbol{\mu}$ and a point $\mathbf{x}_i$ in the dataset. $d_{MA}(\mathbf{x}_i, \boldsymbol{\mu})$ then provides a test statistic to determine whether $\mathbf{x}_i$ is an outlier. This functions well when the distribution of the sample data is Gaussian (for instance) with few outliers, but is not accurate for highly skewed or multimodal distributions, or for sample sets containing a large proportion of contaminated data points (outliers). This last effect occurs because the Mahalanobis distance is not robust — the true centroid and covariance are quickly "masked" by the contaminated data, creating misleading results (Figure 4.1) [45]. One method for dealing with skewed and bimodal data is the k-nearest neighbor approach [46, 47], discussed in Section 4.2.4. In order to mitigate the masking effect, several robust estimators are used in place of $\boldsymbol{\mu}$ and C; one popular choice is the Minimum Covariance Determinant (MCD) estimator, discussed in Section 4.2.3.

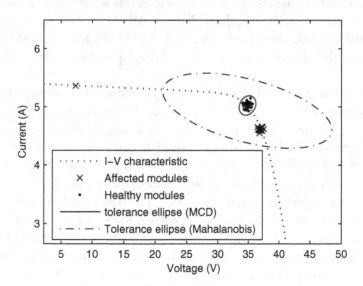

Figure 4.1: Demonstration of masking effect on I-V measurements on an array simulation under ground fault conditions. The 99.7% tolerance ellipse is plotted for each case; the MCD estimator shows little or no masking, while the classical estimator is rendered virtually useless by masking due to multiple clustered outliers.

4.2.3 THE MINIMUM COVARIANCE DETERMINANT (MCD)

The MCD estimator is a multivariate measure of a dataset's center and variation characteristics that possesses good resistance to data contamination by outliers. First introduced by Rousseeuw [48], the MCD seeks to determine the subset of h points whose covariance matrix has the smallest possible determinant. The parameter h is an integer chosen such that $0.5J < h < J$, where J is the sample size. The MCD estimator has a high breakdown point: it can withstand $J - h$ contaminated observations with only finite deviation from the uncontaminated MCD. The MCD estimator is prohibitively expensive to compute for large samples ($J > 100$) and high-dimensional data. However, the FAST-MCD algorithm [45] finds approximate results with acceptable complexity by iteratively modifying the subset of h points such that its covariance determinant decreases with each iteration, rather than testing every possible subset to find the optimal one. If the outputs of the MCD estimator are used in place of $\mathbf{C}$ and $\boldsymbol{\mu}$ in the Mahalanobis distance, a more robust test statistic for outlier detection is produced, as shown in Figure 4.1.

4.2.4 k-NEAREST NEIGHBOR METHODS

The approaches previously mentioned in this section implicitly assume a non-skewed distribution of inputs. Put another way, they assume that the probability density function (Pdf) of the distribution from which observations are taken has elliptical level sets which are all centered around the same point. However, in real applications, the underlying Pdf of the observations may be skewed; in fact, it may be completely unknown. In this case, the k-nearest neighbor method may be used to classify data with only a labeled training set, and without developing any model of the distributions of the classes [49].

The operation of a k-NN classifier is highly intuitive. An arbitrary number of training sets $X, Y, \ldots$ containing vectors $\mathbf{x}_i, \mathbf{y}_i,\ i = 1, 2, \ldots J$ are defined, as is neighborhood size k. Then, an observation $\mathbf{c}$ is presented for classification. Its nearest neighborhood, the set of k training vectors closest to $\mathbf{c}$, is determined using, for instance, one of the distance metrics discussed above. The vectors in the nearest neighborhood are then allowed to vote on the classification of $\mathbf{c}$, and $\mathbf{c}$ is assigned to a class based on which class is most represented in its nearest neighborhood. In a fault detection problem, two classes representing faulty and non-faulty data would be used.

The approach outlined above is effective when high-quality training data for all classes are available. However, in an anomaly detection problem, data is most often available only for the class of normal observations: anomalous events may be extremely rare and may not be clustered into predictable spatial groups. A simple modification of the algorithm presented above transforms the k-nearest neighbor algorithm into an outlier detector. Only one class X is used, representing the non-anomalous observations. A data point $\mathbf{c}$ is once again presented for classification as a normal or anomalous observation and its nearest neighborhood is calculated as before. However, rather than allowing the nearest neighborhood to vote on the class of $\mathbf{c}$, the distance from each of the nearest neighbors to $\mathbf{c}$ is calculated and the sum of these distances is used as test statistic to determine if $\mathbf{c}$ is an outlier. Depending on the application, a fixed number of anomalies may be detected in each pass, or a threshold on distance may be set, above which an observation is determined to be anomalous.

4.3 MACHINE LEARNING METHODS

The field of machine learning is concerned with the development and implementation of computational models which determine their behavior based on an input dataset. Learning machines include feedforward neural networks, support vector machines (SVMs), self-organizing maps (SOMs), and genetic algorithms. Machine learning techniques have found applications in pattern recognition, system control, and data mining, among other areas. This section describes two popular machine learning algorithms, the feed-forward neural network trained with back-propagation and the support vector machine (SVM), and their application in the area of fault detection. The methods presented below require a training dataset containing a wide sample of both faulty and non-faulty data. Such a dataset is not currently available for solar arrays. However, once monitoring systems become more commonplace and data on fault occurrences becomes available, examples of faulty arrays could feasibly be generated using circuit simulators.

4.3.1 FEED-FORWARD NEURAL NETWORKS

The basic feed-forward neural network, also known as a multilayer perceptron (MLP), is arguably the most well-known example of a neural computational model today. In the MLP, simple units known as neurons are connected in a layered topology: each neuron performs a weighted sum of all the neurons in the previous layer and calculates its output based on its activation function $\phi(x)$. In addition to the input and output layers of the network, one or more hidden layers of varying size are used. Arguably, the most common activation function used is the tan-sigmoid function also known as the hyperbolic tangent. The tan-sigmoid function is one of several activation functions which allows the MLP to act as a universal approximator; that is, with a sufficient number of neurons in the hidden layer, the MLP can map any input-output relationship with arbitrary precision. The popularity of the MLP is due in large part to this fact, since earlier neural network models were restricted in the types of functions they could approximate.

Feedforward neural networks are generally trained by some variation of backpropagation. Many approaches to backpropagation exist; their specifics are beyond the scope of this section, but all follow the same basic approach outlined below. During the k-th iteration, a pair of input and output vectors $\mathbf{x}(k)$ and $\mathbf{d}(k)$ are presented to the network for training. The network output $\mathbf{y}(k)$ and error $\mathbf{e}(k) = \mathbf{d}(k) - \mathbf{y}(k)$ are calculated, and the error energy is used as an objective function to optimize network weights using some variant of gradient descent. A more thorough treatment of backpropagation is given in [50].

For the specific case of a classifier for PV arrays, inputs may reflect the current, voltage, temperature of modules or of the array as a whole, measurements of weather data, measurements of grid behavior, etc. Input data may be taken from only the most recent available samples or past samples may be included as their own inputs. The input vector $\mathbf{x}(k)$ may be pre-processed in order to reduce its dimensionality, for instance by principal component analysis (PCA). Outputs may represent the presence or absence of a fault, the type of fault, or its location, among other things. With proper guidance and a good choice of training sets, the MLP is able to map the input vector with information on present and past array status to an output vector giving information on the performance of the array. In the simplest implementation, a neural network would be trained on a two-class dataset, with classes representing faulty and non-faulty arrays.

4.3.2 DRAWBACKS OF NEURAL NETWORKS

Although powerful, there are several notable issues associated with the MLP computational model. In this section, we discuss the existence of locally optimal network weights and issues of network underfitting and overfitting. Arguably, the most serious of these is the existence of local minima in the objective function used to optimize network weights. This implies that gradient descent and related optimization methods are likely to settle on a globally suboptimal set of weights $\mathbf{w}$. Fortunately, in practice many of the local minima are very nearly optimal and the user need not exhaustively search for the global optimum. The problem is also mitigated by the variation of gradient descent step size in an annealing-based approach. Step size is initially chosen to be large, ensuring that the backpropagation fully explores the space of possible weights. It is then decreased, allowing the algorithm to "fine tune" the weights to reflect local details of the error surface.

One problem with the MLP, and with most machine learning paradigms, is the difficulty of generating a detailed, comprehensive and noise-free set of input-output pairs for training. In order to demonstrate other issues encountered in MLP training, a simple example from PV is considered based on the work in [51]. In [51], a neural network is used to approximate the current of a PV array as a function of temperature, irradiance, and load voltage. By holding irradiance and temperature constant, a single-input single-output neural network may be used to predict the I-V curve of the array. A partially shaded array's I-V curve is approximated in order to better show the effects of overfitting and underfitting.

Ideally, the training set contains a large number of vectors throughout the space of all possible inputs. If gaps or low-density regions exist in the training data, the trained MLP is likely to be a poor approximator of the underlying function for input vectors that are not adequately represented in the training set. This is not a problem unique to machine learning — the MLP often generalizes better than more traditional function approximation techniques. However, the MLP is commonly used with very high-dimensional input data and complex non-linear functional relationships, bringing the issue of generalization to greater prominence. Figure 4.2 shows the effect of a gap in training data on the simple example of an I-V curve of a shaded PV array. When a section of the training data is removed, the network is unable to approximate the function over that region.

A closely related issue is optimal sizing of the MLP. If the MLP's weights are not regularized in any way or it contains too many neurons, overfitting is virtually guaranteed to occur. On the other hand, if too few neurons or overly agressive regularization is used, underfitting is the result. As intuition would suggest, overfitting occurs when the network has too many degrees of freedom and

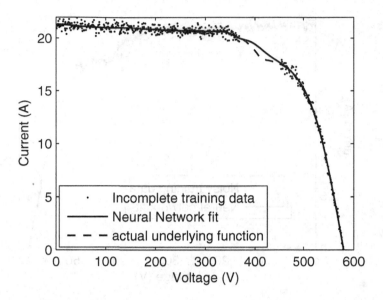

Figure 4.2: Effect of gaps in training data on neural network training.

approximates a functional relationship between input noise and output, when in reality none exists. Underfitting, on the other hand, occurs when the network has too few degrees of freedom and is therefore unable to accurately approximate the underlying function. Figure 4.3 shows the effect of overfitting on the approximation of a function in noisy training data.

Although potentially serious, these issues do not prevent the MLP from being a useful tool. Performance is usually not extremely sensitive to network size or regularization parameters and a wide range of network topologies may produce good results.

4.3.3   SUPPORT VECTOR MACHINES

Another highly successful machine learning model is the support vector machine, or SVM. The SVM is similar to the feed-forward neural network in that its operation is defined by weight matrices which determine the input-output mapping. Unlike the multilayer perceptron, however, the SVM's performance metric is a convex function. As such, the uniquely globally optimal weights for an SVM are relatively easily determined using convex optimization techniques. Like the multi-layer perceptron, the SVM may be used as a classifier or in order to approximate a function. However, the vast majority of successful SVM applications have been in classification problems; this application is treated below.

The SVM classifier operates on the principle of finding the optimal separating hyperplane between two datasets. This optimal hyperplane is defined as the plane which maximizes the margin

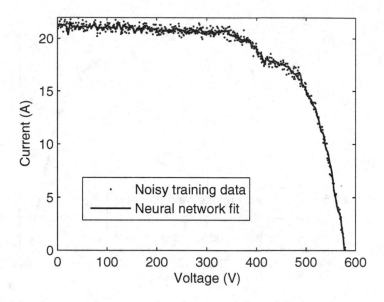

Figure 4.3: Overfitting due to large network size with no regularization.

of separation between the two classes. The points at which this minimum distance is achieved are known as support vectors.

The above-mentioned definition of the optimal separating hyperplane is, of course, meaningful only for two linearly separable classes. Vapnik et al. [52] performed extensive work in generalizing the SVM to non-separable datasets and non-linear boundaries. The problem of optimally dividing two non-separable classes is approached by assigning a "soft margin" which penalizes the misclassification of data while still maximizing the margin of separation between the correctly classified data. Nonlinear class boundaries are introduced by expanding input vectors to a higher dimensionality using a nonlinear kernel function. By appropriately choosing the SVM kernel, arbitrarily complex classification boundaries may be defined and the SVM becomes a universal approximator in the same sense as the MLP introduced in Section 4.3.1.

4.4   SUMMARY

Anomaly detection is a well-studied field which may be applied to the detection of faults in PV arrays. The non-independent behavior of PV arrays under fault conditions creates unique challenges and requires a more sophisticated approach than traditional non-robust statistical tests. Methods from the fields of robust statistics and machine learning may be effectively adapted to the problem. While the machine learning-based methods presented above require a training dataset with extensive examples of normally operating and faulty arrays, the robust statistics-based approach does not require training data. However, as more arrays are equipped with monitoring systems the development of a strong training dataset is expected to become more feasible. Both classes of anomaly detection techniques have the potential to increase PV array uptime and power output and decrease the time between the occurrence of a fault and its repair.

CHAPTER 5

Array Topology Optimization

This chapter discusses reconfiguration of the electrical topology of PV arrays for maximizing array energy production. PV module currents and voltages are not sufficient to operate an inverter. In order to meet the specifications of the inverter, PV modules are connected in series and parallel to increase the voltages and currents respectively. The behavior of series connected PV module strings and the interactions between the strings affects the power output by the array (Section 2.3). The efficiency of operation of a solar array depends not just on the weather conditions but also on the array topology [53, 54]. Conventionally, fixed PV array system topologies were determined taking into consideration the weather condition for the entire year. This design does not provide the maximum power levels for a given day or season. A fixed topology is unable to bypass failed or under-performing modules, reducing the overall array output. To improve the array output power, reconfigurable systems that can change their topology need to be developed. This chapter gives a brief overview of the design methodologies for fixed topology PV arrays and discusses the dynamic reconfiguration of array topology.

5.1   FIXED ARRAY TOPOLOGY DESIGN

The design rules for the conventional fixed topology array need to be understood as these set up the constraints for the reconfigurable topology system design. The optimal number of modules in series and parallel for a PV array design depends on several factors such as the maximum and minimum temperatures in a year at that site, the operating range of the inverter, the wiring used and the nature of the array (standalone or grid tied). This section describes the methods used to determine the number of modules in series and parallel for a fixed topology array. The feasible topologies for the reconfigurable array are all the topologies that satisfy the fixed topology design criteria.

5.1.1   SIZING STAND-ALONE PV ARRAYS

The IEEE standard 1562 [55] provides the guidelines to fix the array size for stand-alone PV arrays. The following constraints are imposed for the design standard: the PV array is used to charge a battery and is the only charging source. The PV array is sized to replace the energy consumed from the battery by the load (after accounting for system losses and inefficiencies of the module). The effects of shading are not considered. The solar radiation for the worst case month (minimum irradiance) is assumed. Also, worst case is assumed for the monthly load (maximum load).

　　　The number of series connected modules is obtained as the ratio of the nominal system voltage and the nominal module voltage rounded up to the nearest whole number. PV modules were traditionally designed to charge batteries. Here, the nominal module voltage is the battery voltage best suited for the given module while the nominal system voltage is the battery's voltage. The number of parallel strings N_P is obtained as a function of parameters such as the average daily load, the array to load ratio (ratio of the average daily PV energy output to that of the average daily load), the loss in the system, the current at maximum power point, and sun hours (the number of hours of sun at standard irradiance of 1000 W/m^2 per day that would equal the amount of solar radiation received). IEEE 1562 provides information on determining the values of these parameters for designing a stand-alone PV system.

5.1.2   SIZING GRID CONNECTED PV ARRAYS

The design criteria for grid connected PV systems is specified in [56]. Inverters have a fixed range of operation in terms of minimum and maximum voltage for the DC input. The operating voltage range of the inverter limits the number of modules in a string. The maximum voltage generated by the array must not exceed the inverter's maximum input voltage under any condition. As the module voltage increases at lower temperature, the open-circuit voltage of the array must not exceed the maximum operating voltage of the inverter on the coldest day of the year. The maximum number of modules in a string is given by the ratio of the maximum input voltage of the inverter and the open-circuit voltage of a module at the lowest winter temperature of the year

The minimum number of modules in a series string is determined by the minimum input voltage requirement of the inverter and the maximum temperature at which the modules need to operate. It is given by the ratio of the minimum input voltage of the inverter at maximum power point (MPP) and the MPP voltage of a module at the highest module temperature during the year.

The limitation on the number of strings in parallel is determined by the maximum current that can be input to the inverter as well as the current carrying capacity of the wires used. The maximum number of strings in parallel is given by the ratio of the maximum current that can be input to the inverter and the short-circuit current at maximum irradiance for the given string.

In the U.S., National Electric Code regulates the tolerance levels for the current carrying wires used in PV arrays. The code requires that the wires used be rated at at least 156.25% of the maximum short-circuit current they might be expected to carry. This places an additional limitation on the maximum number of strings that can be put in parallel.

5.2   NEED FOR RECONFIGURABLE TOPOLOGY

PV modules in general deviate from their rated output power and modules in the same PV array differ in the currents and voltages produced. This mismatch in the modules leads to non-optimal performance of the array. The current in a string can be affected by a single under-performing module. This mismatch is much more disruptive when one or more modules are shaded or faulty as seen in Chapter 3. The array topology can at times be altered to minimize the effect of these faults. It is experimentally shown in [57] that changing the traditional series-parallel (SP) configuration to alternate topologies such as bridge link (BL) and total cross-tied (TCT) configurations (2.3) results in a 4% increase in the array power under shading conditions. It is possible to improve the number of hours an inverter operates in a day by using dynamic reconfiguration. This section provides examples of both these benefits.

5.2.1   INCREASING INVERTER UPTIME

When the voltage of the array is below the threshold voltage of the inverter, it ceases to operate. Reconfiguration can be done to increase the amount of time during the solar day that the array can be operated. More modules can be put in series early morning and late evenings when the irradiance is low to increase the voltage above the inverter's threshold.

Figures 5.1 and 5.2 show the switching scenario for an array of 44 modules connected to an inverter with an operating range of 300–600 Volts. The vertical axis shows the DC power output by the array in Watts and the horizontal axis shows the DC voltage generated by the array. The vertical line marked "Inverter Threshold" marks the input DC voltage below which the inverter will shut down. The best case scenario for power generation would involve the array configuration that produces the maximum power while remaining above the voltage threshold of the inverter.

Figure 5.1 shows the power output for two specific configurations of the 44 module array when the irradiance is 700 W/m^2 and Figure 5.2 shows the power output for the same two configurations when the irradiance is 100 W/m^2. Both these configurations are feasible topologies, i.e., they satisfy the criteria for number of modules in series and parallel given in Section 5.1.2. At higher irradiance values, the configuration with 11 in series and 4 in parallel results in all the modules being utilized, producing a higher output power. This will be the desired configuration at higher irradiances. At lower irradiance, this configuration falls below the inverter's operating voltage range and cannot produce any output. Switching to a configuration with more modules in series pushes the voltage levels above the threshold of the inverter as seen in Figure 5.2. Therefore, at lower irradiance values, the 14 in series and 3 in parallel is the desired configuration. The simulations were performed assuming no mismatch between modules and module cell temperature as 45°C. The PV module was assumed to be Sharp NT-175U1 and the Sandia performance model discussed in Section 2.2 was used to calculate the module output for the different irradiance values.

5.2.2   IMPROVING ARRAY OUTPUT

Any under-performing module in a string affects the output of the entire string (Chapter 3). It is sometimes advantageous to rearrange the electrical configuration of the array to either remove the under-performing module entirely or position it in a way that its detrimental effects are reduced.

Figure 5.3 illustrates a scenario where reconfiguration is beneficial. The array is assumed to be composed of 52 modules in a 13-series, 4-parallel strings configuration. Two of the 52 modules are shaded for a certain duration of the simulation. This is achieved in the simulation by reducing the irradiance values for the two modules to 25% of the actual irradiance. It is seen that by removing the two shaded modules from the array and configuring the remaining 50 modules in 10 series, 5 parallel configuration, we get an increase in power output. This is due to the fact that the two shaded

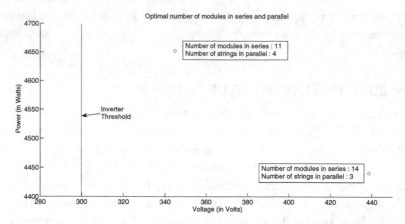

Figure 5.1:   Effect of irradiance on array topology for irradiance 700 W/m^2.

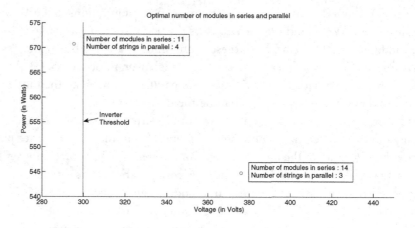

Figure 5.2:   Effect of irradiance on array topology for irradiance 100 W/m^2.

modules in the 13-series, 4-parallel case are bypassed and the remaining modules in the strings with shaded modules operate at a suboptimal point to compensate for the shaded modules (Section 3.1). This results in lower power output compared to the 10 series, 5 parallel case where all the modules are operating at near maximum power. Once the array returns to the normal mode of operation (without shading), the 13-series, 4-parallel combination which utilizes all 52 modules performs better. The simulations were done using the five-parameter model (Section 2.2) implemented in SPICE.

The array cannot be electrically modified to any arbitrary topology. The possible topologies are restricted by the design requirements presented in Section 5.1.2. In the example given above,

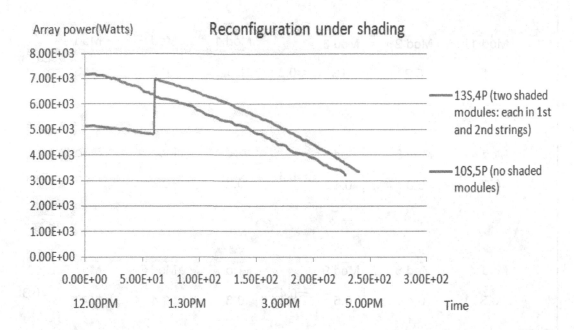

Figure 5.3: Array output power for two different topologies when two modules are shaded for a brief period during the day in a 52 module array. During the period the modules were shaded, the 10 series, 5-parallel topology performed better than the 13-series, 4-parallel topology. When there was no shading, the 13-series, 4-parallel topology performed better.

the inverter used was assumed to be a Satcon PowerGate Plus 50kW inverter. This determined the minimum and maximum inverter input voltages (300 and 600 Volts, respectively). Maximum and minimum temperature values were determined based on meteorological data for Phoenix, AZ. These values were used to determine which topologies can be safely used, and the topology producing the maximum output among those was chosen.

5.3   EXISTING RECONFIGURATION METHODS

Fixed array topology designs are suited for year-round operation of PV arrays under ideal conditions. However, as seen in Section 5.2, array power output is reduced when there are faulty or shaded modules within the array. Reconfigurable topologies can improve the array performance under non-ideal conditions while retaining the benefits of fixed topology design. This section describes the existing reconfiguration methodologies.

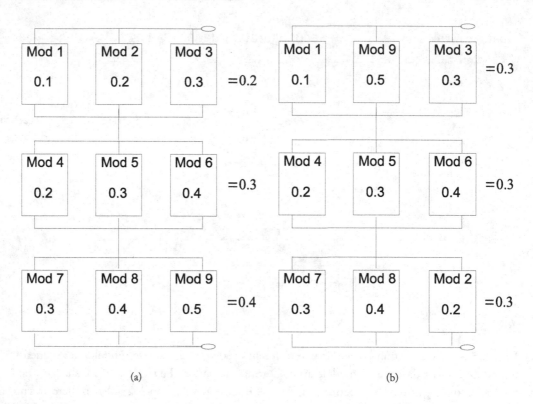

(a)   (b)

Figure 5.4: Irradiance equalization. The mismatched modules are rearranged such that each parallel block has equivalent irradiance before being connected in series. This results in better matching of the currents of each of the parallel blocks. This ensures that the output power for the second configuration is better than that of the first.

5.3.1   IRRADIANCE EQUALIZATION

Methods exist in literature that reconfigure or modify the array topologies to reduce the detrimental effects of shading. The irradiance equalization method [58] is applied to PV arrays that are connected in total cross-tied (TCT) configuration (Section 2.3). Depending on the irradiance received by the modules, they are relocated such that each parallel block of modules has similar effective irradiance before being connected in series. The relocation of the modules is achieved by employing a switching matrix.

The presence of a shaded module in a string limits the current in the string to the current of the shaded module. Connecting PV modules with similar operating characteristics in a string would then increase the overall power output. The concept of irradiance equalization as demonstrated by [58] is shown in Figure 5.4. The PV array consists of 9 modules connected in TCT configuration. Each row in Figure 5.4 consists of three modules connected in parallel and can together be considered as a single bigger block module. The current generated by the block module depends on the irradiance received by the individual modules. Three such block modules are connected in series. The effect of output current of each module is calculated as the effective average irradiance using the diode equations described in Section 2.1.

The effective average irradiance of each block of parallel connected modules is the average of the individual effective irradiances. Figure 5.4(a) shows block modules with unmatched average effective irradiance values. The switching matrix is then used to rearrange the electrical location of the modules 2 and 9 and the irradiance matched configuration is shown in Figure 5.4(b). The output for configuration (b) was found to be higher. This is because the current mismatch between the blocks is reduced before connecting them in series.

The requirement that any module should be electrically relocatable to any location within the array increases the complexity associated with the switching matrix as the array size increases. For very large arrays, the excess wiring and switching matrix requirements would make it impractical to implement this method. The complexity can be reduced if only some of the modules in the array can be switched while others are electrically stationary. This approach is investigated in the adaptive banking method described in Section 5.3.2.

5.3.2   ADAPTIVE BANKING

The adaptive banking method [59, 60] reconfigures the PV array to provide maximum power output under different shading conditions. Consider the array of 16 modules connected in the TCT configuration given in Figure 5.5(a). Under normal operating conditions there is no need for any changes to this configuration. However, three of the modules in the array are currently severely shaded, turning the bypass diodes on. The maximum current that can be generated by this array is now limited by the first row which effectively has only two modules. This detrimental effect can be minimized to a certain extent by splitting the array into two parts: the fixed part and the adaptive part. Here, the first three columns can be made the fixed part and the last column can be made the adaptive part. Each module in the adaptive part can be connected to any of the rows of the fixed part. This is accomplished by employing a switching matrix constructed using relays or electrical switches. With this arrangement, the severely mismatched condition of rows in Figure 5.5(a) can be rectified, as shown in Figure 5.5(b). Here, two modules from the adaptive part of the array are added in parallel to the first row, which has 2 modules shaded, and one module is added to the second row, which has one module shaded. Now, each row in the array has at least three non-bypassed modules, therefore the current is not restricted to two modules as in the previous case. If the mismatch is not severe, the most illuminated solar module from the adaptive bank is connected in parallel to the row of the fixed part that has the least power output (the most shaded row).

This method is less complex than the irradiance equalization method at the cost of reduced versatility. This might result in the best possible configuration not utilizing some of the modules

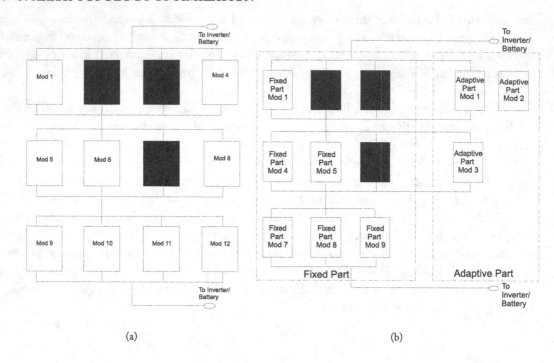

(a)   (b)

Figure 5.5: Adaptive banking method. There is mismatch caused in the array due to the shaded modules in configuration (a). This is overcome by making the last column of the array reconfigurable as shown in configuration (b).

from the adaptive bank, resulting in reduced power output compared to the irradiance equalization case.

5.3.3 OTHER METHODS

Tria et al. [61] described an algorithm to reconfigure the PV array for maximum power transfer. Voltage matching is done by optimizing the number of series connected modules until the quotient of the measured array voltage divided by the number of modules in series approximates the voltage at maximum power of the module V_{MP}. Once the voltage is matched, currents are matched in the same manner by adding strings in parallel depending on the availability of modules. Simulation results show an increase in output power in the reconfigured array as compared to the static case with the amount of increase depending on the type of load.

The reconfiguration of a solar array for use in outer space applications is the basis of a U.S. patent [62]. As the spacecraft containing the solar array moves away from the sun, the current decreases and voltage increases due to lesser irradiance and cell temperature, respectively. The array here consists of at least four strings and is reconfigured to put more modules in parallel to improve the current generated as the space craft pulls away from the sun.

5.4 REQUIREMENTS FOR RECONFIGURABLE PV SYSTEMS

Reconfiguration of topologies requires a means to electrically modify the connections within an array. This is typically achieved by means of a switching matrix arrangement. There must be ways to control the operation of the switching matrix through commands issued from a device such as a micro-controller.

There must be automated algorithms that can determine when reconfiguration needs to occur and calculate the new topology. Algorithms are needed to distinguish healthy modules from under-performing modules which can be removed, if necessary, from the new topology. Algorithms to determine faulty or under-performing modules were discussed in detail in Chapter 4. The array cannot be reconfigured to a new improved topology instantaneously. For safety reasons, the inverter needs to be shut down before reconfiguration and started up afterwards. The loss of power due to the shutting down of the inverter must be included in the calculations determining whether reconfiguration is required. The algorithms must have the ability to control the inverter operation to turn it on/off.

Algorithms to reconfigure the topology need accurate information about the status of the array. This requires a monitoring system that can determine voltages and currents of individual modules, irradiance and ambient temperature values at the site of operation and transmit this information to the systems running the algorithms. The monitoring system must collect information in a periodic manner and store it for later use by technicians. A more detailed study of monitoring systems for PV arrays is presented in Chapter 6.

5.5 SUMMARY

Traditional fixed topology designs are unable to adapt to faults in the array and mismatches between modules. Reconfigurable topologies rearrange or bypass modules within the array and reduce the mismatches between modules. They can also improve the inverter uptime. Reconfiguration methods, such as irradiance equalization and adaptive banking, show the advantages of reconfigurable topology under mismatch or shading conditions. Dynamic reconfiguration is only possible when the array has certain additional capabilities such as switching matrices and automated monitoring and control.

CHAPTER 6

Monitoring of PV Systems

This chapter explains the need for an automated PV array monitoring system, some existing methods of PV monitoring, and discusses the different sub-systems required for DC side monitoring. Because of the requirements of space and environmental conditions, many large-scale PV arrays are located at remote destinations where access to them is difficult. There is also a growing trend of smaller arrays installed on rooftops of individual houses or commercial buildings. Although they may not be in remote locations, access to such systems is limited as well. Faults such as ground faults or shading discussed in Chapter 3 often go unnoticed for long periods of time resulting in significant reduction in array power output. Technicians sent to rectify the faults have to spend a significant amount of time determining the location and nature of the fault manually. Hence, there is a need for low-cost monitoring systems to automatically obtain information about the performance of the array and detect faults.

6.1 THE NEED FOR AC SIDE MONITORING

PV arrays require effective monitoring on the AC side of the inverter to help the utility company maintain power quality, obtain information about outages, and adhere to standards for connecting PV to the grid.

6.1.1 PREVENT OUTAGES

The power output of PV systems, unlike traditional fossil fuel-based generation schemes, depends on factors such as solar irradiance, cloud cover, and temperature [3]. The power output of the array might drop or increase drastically at any instant in time depending on external factors. Sudden drops in output power from large PV plants could lead to power outages [63]. As PV achieves higher grid penetration, a centralized monitoring system becomes essential. Such a system would allow greater control over the contribution of specific PV plant to the grid. If the PV plant with a monitoring system has reduced output, controllers would be able to supplement the power with traditional systems such as a storage battery or a diesel electric plant.

6.1.2 ISLANDING PROTECTION

The addition of arrays to the power grid poses a safety risk known as islanding. Islanding occurs when a fault in the wider electrical grid causes a power outage, but a PV array does not turn off. This might result in the PV array supplying the local loads resulting in the formation of a small "island" in the grid. This island of energized wires poses a threat to repair technicians seeking to fix the fault. Modern inverters are, for the most part, equipped with anti-islanding features which detect when the AC grid goes down and shut off the inverter. In addition to this safeguard, many electrical companies require that residential arrays connected to their grid come equipped with a manual disconnect switch, to be opened by a power company technician before beginning work on nearby parts of the grid.

The IEEE standard 1547 [63] provides guidelines for interconnecting a distributed generation resource (such as PV) with power systems. The standard provides guidelines such as bounds on voltage and frequency distortion while laying out response requirements to abnormalities. The guidelines mandate that the PV array interconnection system must detect unintentional islanding and cease to energize the electric power system. The standard mandates that if the aggregate of the distributed resources at a single point of common coupling (point where a local electric power system is connected to an area electric power system) equals or exceeds 250 kVA, there must be provisions to monitor the connection status, power output and voltage at the point of distributed resource connection.

6.1.3   IMPROVE POWER QUALITY

Power utility companies are responsible for ensuring power quality. Significant deviation in the magnitude, frequency, or purity of waveforms results in power quality problems. The utility has no control over the amount of current a load might draw. Power quality is maintained by keeping the voltage and frequency within certain limits. The output of PV arrays vary significantly over time resulting in potential power quality problems. Monitoring and control of PV arrays can be used improve power quality by changing the outputs such as voltage and frequency of a PV array.

The availability of monitoring information from various distributed sources in a given area can provide better insight for the electric power company. This information provides avenues for better understanding of a given area and means to improve power quality. As more PV is added to the grid, power quality becomes a serious concern.

6.1.4   PERFORMANCE EVALUATION

Effective monitoring of a PV array can be used to evaluate the performance of the array. For instance, the voltage and currents generated by the array can be used to determine the annual energy production. This can then be used to determine the cost per watt of the PV system. The electrical outputs such as power generated can be compared to the weather data using PV models to determine the efficiency of operation of the system.

6.2   THE NEED FOR DC SIDE MONITORING

Monitoring on the AC side of the array cannot be used to determine the location of the faults, if any, within a PV array. PV arrays require monitoring on the DC side to identify faulty modules, calculate the efficiency of operation of the modules and determine long-term and short-term trends of the system.

6.2.1   IDENTIFY FAULTS

Methods of fault detection that rely only on human operators are not accurate and might result in faults remaining unnoticed. Automated monitoring systems must be used to identify the presence of faults in an array or sub-array. Monitoring on the AC side cannot detect faults that affect only a few modules in the array. The loss of power due to such a fault is indistinguishable from measurement errors and seasonal variations. Hence, monitoring is required on the DC side to detect faults effectively. For instance, [44] suggests an ad-hoc method that utilizes information obtained at the end of each string to identify faults that reduce the string output by over 20%. Automated monitoring systems can use sophisticated signal processing algorithms to identify the exact location of the fault within the array. A signal processing algorithm that utilizes measurements from each module to detect ground fault and arc fault locations at over 99% accuracy is described in [64].

6.2.2   EVALUATING TRENDS

The outputs collected on the DC side can be compared to the historical data to determine the long term trends. For instance, module data collected from previous years can be compared with that of the current year to calculate the module degradation factor. The output of the modules can be compared to weather data to update PV models.

6.3   EXISTING METHODS OF MONITORING

Several systems have already been developed to monitor PV arrays and modules. We look into some of these systems currently being used, emphasising systems that perform intra-array monitoring. The usability of each system varies depending on their targeted capabilities.

A monitoring system used to evaluate the performance of PV array installed on a building is given in [65]. The monitoring system measures voltage, current, and power at the AC output of the inverter. It measures the solar intensity using two photo diode sensors and taking their average. Module internal temperatures are obtained using temperature sensors. The measurements are done over 20-min intervals. The data can be used for both short-term purposes such as monitoring the system's status, and long-term purposes such as monitoring the deterioration of components. Comparison of the different strings helps evaluate the effect of shading on the array. The correlation of output power with temperature can be used to determine the effect of module temperature on output power for the same irradiance. These measurements can be used to evaluate the annual energy production and cost of electricity produced by the array. Measurements are transferred to a computer enabling internet access to the data. The authors mention the use of commercially available data acquisition systems to transfer the data to the computer. Here, the monitoring system does not consider module level measurements and communication systems for such an arrangement.

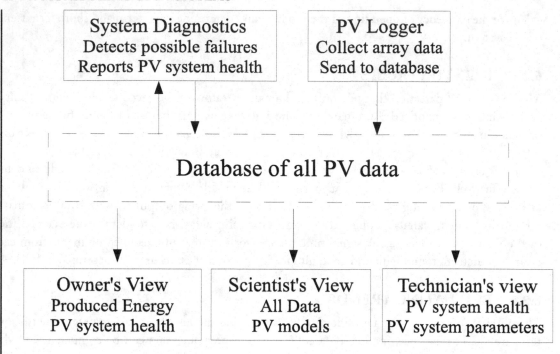

Figure 6.1: Schematic for data access according to the need of each user.

Kolodenny et al. [66] proposed an approach that uses modern informatics tools such as XML to analyze the acquired data. Their goal is to analyze the performance of a PV system of any type and size. They propose a protocol called PV mark up language (developed from XML) to automatically access, extract, and use data from several sensors systems/database sources. The system collects and classifies the data. Depending on whether the user is a technician or plant owner or scientist, different views are presented containing only the most relevant information. Figure 6.1 shows the schematic for their data access system. A PV logger system constantly collects information about the state of the array (such as voltages, currents and irradiance) and updates it in a database. The system diagnostics retrieves the data and detects possible failures. This information is then updated in the database. The different users can query the database and obtain information they are interested in. The owner can view the overall system health and output of the array while the technician can view the PV system parameters. The scientist's view provides access to all the data collected including the PV system models. This work provides a comprehensive method to store and retrieve PV array data and can be used with different current voltage sensor location and weather data. However, it does not describe a complete PV monitoring system which includes sensing, data communication, and user interface systems.

A simple and cost effective method of monitoring a PV power station using a GUI built in

NI LabVIEW is presented in [67]. The set-up consists of current and voltage sensors connected to a micro-controller through an analog multiplexer. Also included are sensors for measuring the module surface temperature and irradiance. The collected data can be used for both monitoring and control. The micro-controller is interfaced to a laptop through a serial port where the data can be viewed and analysed using a GUI designed using LabVIEW.

6.4   MONITORING SYSTEM CONSIDERATIONS

Each monitoring system discussed in the previous section addresses specific aspects of PV array monitoring such as data acquisition, data management, and graphical user interface. These different aspects of a PV monitoring have to be combined in a single automated system.

The block diagram for a PV array with an automated monitoring system is shown in Figure 6.2. The PV modules are connected to the inverter to form the PV array. Sensors are placed to collect the PV currents and voltages from the DC input to inverter, end of each string and/or individual modules. The measured parameters are transmitted to the central server through a communication channel. The communication channels used can be ethernet, wireless, or power line. A weather station records the irradiance and temperature values and transmits them to the central server. The central server maintains a database of the array outputs and weather data. It runs PV models to calculate expected output values and fault detection algorithms to detect the presence of faults in the array. The array operator has access to the expected and actual output and fault information through an easy to use GUI.

There are several aspects to be considered for an effective monitoring and control system for PV arrays. These include the parameters to be measured, the placement of sensors for measuring these parameters, communication systems that can transfer the measurements to a central database, algorithms to manage and process the collected data, and user interfaces to visualize the state of the PV arrays. This section discusses these aspects briefly.

6.4.1   SENSOR MEASUREMENTS

PV array characteristics can be determined from the measured parameters such as voltage, current, and irradiance. The measurements can be made at different locations in an array, from the inverter to individual modules.

Inverter Level Measurements

Most currently used PV monitoring systems record current and voltage measurements at the inverter to analyze the performance of the array. Using the data for a typical year at the site and models such as the Sandia performance model [3], the expected output power of the array can be calculated. These can be compared with the actual power output by the inverter to determine whether the array is operating without any faults. For example, using monthly averaged energy at the inverter as a metric, loss of energy in the array was identified and rectified in [3]. Inverter faults are characterized by large variation between expected and actual AC power while actual and expected DC power

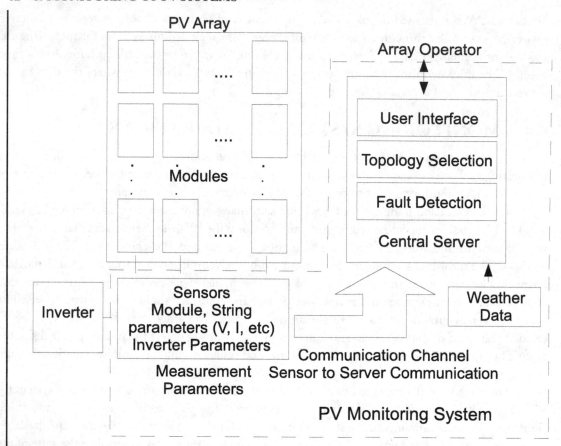

Figure 6.2: Block diagram of a PV array with automated monitoring system.

remain similar. Array faults are characterized by significant variation in expected and actual DC power.

String/Module Level Measurement

Although measurements at the inverter level enable us to identify the overall health of the system, they cannot in most cases detect non-catastrophic faults. They do not provide enough information to identify the location and nature of faults immediately. Identifying and correcting a fault or under-performing component in the array still involves taking field measurements by technicians. Higher-resolution data acquired by means of additional sensors installed on individual modules or strings provides more precise understanding of the array. These sensors can measure module or string currents and voltages. These additional readings help to identify the modules that need to be inspected in case of a fault.

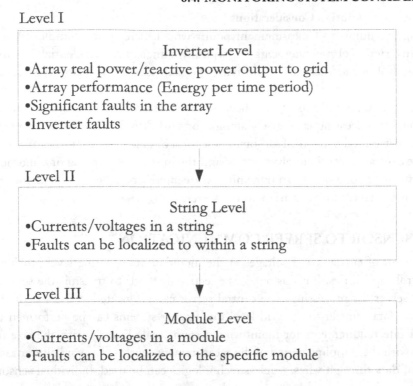

Level I

Inverter Level
•Array real power/reactive power output to grid
•Array performance (Energy per time period)
•Significant faults in the array
•Inverter faults

Level II

String Level
•Currents/voltages in a string
•Faults can be localized to within a string

Level III

Module Level
•Currents/voltages in a module
•Faults can be localized to the specific module

Figure 6.3: Levels of measurement in a PV array starting at the level of the inverter and going up to the level of individual PV modules.

It is possible to measure the temperature, wind-speed and irradiance at the installation site. Approximate estimates of the module cell temperature can be estimated from the atmospheric temperature and wind speed [3]. These can be used in models described in Section 2.2 to predict the output of the PV array. Significant differences between the predicted outputs and actual measurements can be utilized in an automated fault detection set-up.

The measurements at string level provide a more definitive description of the performance of an array when compared to only inverter level measurements. However, they are far from perfect. Cell temperatures calculated from atmospheric temperature do not reflect the exact temperatures of the cells required for use in the PV models. This can be due to manufacturing process variations, wiring effects and other factors such as high resistance connections. Even cells within a single module might have variation in their temperatures. Better estimates of cell temperature may be acquired by sensors on the back surface of modules. These module level measurements can then be used in conjunction with the string level and inverter level measurements. The different measurement levels for a PV array and their capabilities are summarized in Figure 6.3. As the number of measurement points increases, the automatic monitoring system becomes more effective in localizing the faults.

Measurement Resolution Considerations

Increasing the number of measurement points would increase the complexity of the monitoring system. Inverter level measurements are typically sufficient for residential PV arrays which have only a handful of modules since identifying the fault location is relatively straightforward. Large scale PV arrays with thousands of modules would require module/string level measurements. Since there is just a single point of data collection for inverter level monitoring, transmitting the data to a server for processing is relatively straightforward. Monitoring at the string and module level requires a distributed network of sensors. These must be able to transmit the information collected to either the central server themselves or transmit the information among one another until it reaches a gateway sensor node which can transmit to the central server. Section 6.4.2 discusses technologies to transfer the data from the sensors to the server.

6.4.2   SENSOR TO SERVER COMMUNICATION

Measurement of currents and voltages at the module level or string level provides better analysis and control capabilities. But this requires effective methods to transmit the sensor information to a central server for processing. The central server can be located at the site of the PV array and analysis or data transfer to the grid level monitoring systems can be performed from this server. The data rate requirements for monitoring systems scale linearly with the size of the array, and inversely with the sampling period of the measurement sensors. To avoid clutter associated with the transfer of data through wires, wireless technologies can be used. Individual sensors would require relatively low data bandwidth to transfer the measurements to the server. The power consumption for the data transfer must be low to enable extended use of the sensors in the field without battery replacements. The requirements of low power and low data rate makes wireless technologies such as variants of IEEE 802.15.4.2006 (Zigbee, WirelessHART) and in some cases Ultra-wideband (UWB) practical. Data communication for PV systems to transfer the data from the sensor can also utilize power line communications (PLC) which would utilize the existing infrastructure to transfer the data across power lines to the inverter. Any data transfer mechanism can be used to transmit this data collected at a single point to the server. A brief description of these two types of communication systems is given below.

Power Line Communications

To avoid extra wiring, or interference with existing devices, PLC technologies [68, 69] can be used for communications. PLC is widely used both over high-voltage and low-voltage lines. However, there is no adopted IEEE standard, especially in the context of PV arrays. Communication of data on the DC generation side of the PV array can make use of protocols such as RS 232 [70]. The PLC channel introduces large amounts of time and frequency variation in the medium of propagation. The time-variation of the communication medium is due to sudden electrical load changes which alter the overall impedance of the system. The frequency variation is due to the dispersive nature of the medium which causes reflections of the transmitted waveform.

Wireless Technologies for Communication

Using wireless technologies to transmit the data from the sensors to the server can be considered as an alternative to PLC. ZigBee [71] is a wireless standard which is an extension of the IEEE 802.15.4.2006 standard focusing on low-cost, low-power communication between devices located within a typical range of 10–75 m. The data rates are in the order of kilobits per second, which is sufficient to transmit the array state to the server or the gateway sensor. ZigBee utilizes significantly lower power compared to standards such as Wi-Fi and Bluetooth.

Ultra wide band (UWB) communication can be used if the distance between the transmitter and receiver is extremely small (around 10 m). UWB communications consume low power and transmit information without affecting other systems using the same frequency band. However, UWB can handle much higher bandwidth than ZigBee. UWB communications do not have any security protocols whereas ZigBee uses encryption.

These wireless technologies for data communication of the PV system suffer from some of the same problems as that of PLC. These might include interference from other wireless devices, signal distortions in the form of time-dispersion and signal fading.

There is no standard that dictates the usage of these technologies in the PV area. It is advantageous to have PLC communication over ZigBee when the data rates are low (in the order of hundreds of bits per second) and the coverage requirements of the network are more than the range of ZigBee (say over 100 m). ZigBee is better suited for data rates in the order of kilobits per second and a smaller coverage area [72, 73].

6.4.3 USER INTERFACE

Once the sensors and the communication systems to the server are set up, there is a need for algorithms and user interfaces that enable the user to make sense of the data and take appropriate actions. Simple and easy to use GUIs need to be developed for visualizing solar array data. Such GUIs can provide information on overall health of the system. They are linked to sophisticated data processing algorithms and models of PV systems and can be used to evaluate the PV array performance, detect failures etc.

The GUI must provide a real time update of the current state of operation of the array. This includes the AC/DC power output generated by the array, faults if any detected and the status of the battery. It can include other information such as real and reactive power output to the grid, total power output during the day/week/month and the efficiency of array operation. The cost of power generated can be displayed. This is especially useful for residential PV arrays as owners can keep track of the effective amount earned from their array after energy consumption for their residence.

The GUI can provide comprehensive tools to analyze the array as a whole or individual modules, strings, etc. Fault detection algorithms can be used to display the nature and location of any faulty modules. The GUI can provide tools to plot/analyze the array and module output against weather data such as irradiance. This can be done for current day or across several days using stored data and can be used to study the short term and long term trends in the array.

The GUI should be connected to other systems such as audible alarms to alert the operator when faults occur. They can be connected to emergency systems such as fire or flood alert systems as required. They can be integrated with perimeter security systems such as video cameras to provide a single point of reference for the human operator.

If required, a PV monitoring system GUI must be able to enforce data security based on access privileges. It can be used to generate automated reports regarding the performance of the array. The GUI must, if required, be capable of handling multiple arrays within a site.

6.5 SUMMARY

This chapter discussed the need for monitoring systems for photovoltaic arrays. Such monitoring systems should be able to provide the system operator with information regarding the array's operation. The level of detail collected from a PV array determines the accuracy and capabilities of the monitoring system. Small scale PV arrays generally do not require high resolution data collection whereas large scale arrays do. The data collected from the sensors must be transmitted to a central server for processing using wireless or power line channels. Wireless channels are suited for higher data rate and smaller coverage areas while power line communication can handle large coverage areas at lower data rates. The central server must contain algorithms used for processing the data collected from the array and a user interface that lets the operator visualize and take actions based on the data.

CHAPTER 7

Summary

A "smart PV array" may be equipped with monitoring and control capabilities in order to achieve an increased operating efficiency and fault tolerance. Signal processing and pattern recognition techniques are used to monitor photovoltaic arrays and detect and respond to faults with minimal human involvement. When suitable electrical capabilities for switching are installed, the electrical topology of an array can be dynamically optimized to mitigate the effects of module failures and ensure operation of the array at the highest possible efficiency. In this book, a brief overview of photovoltaic module and array behavior was discussed, followed by signal processing and communication work in the following domains: fault detection, array reconfiguration, and array monitoring.

Chapter 2 described the operation and electrical behavior of Photovoltaic cells, modules, and arrays. The effect of irradiance and temperature on the performance of a solar cell was studied. Increased solar irradiance primarily affects the current output of PV cells, while increased temperature decreases available voltage through its effect on the diode behavior of a PV cell. Two models for predicting PV module behavior, the Sandia and five-parameter models, were described and compared. Three electrical topologies used in PV plants — the S-P, TCT, and BL configurations — were presented and their electrical characteristics under ideal conditions were explained.

Chapter 3 discussed the types of faults that occur in a PV array. Some conditions, such as shading, soiling, and mismatch, can significantly reduce power output but are not hazardous. Current practice is to allow these conditions to continue indefinitely. Other conditions, such as ground and arc faults, are hazardous and must be repaired, but there is typically no reliable method to remotely detect a fault location. It is observed that faults occurring in individual modules have a significant influence on the array efficiency: the performance of PV modules is highly interdependent.

Chapter 4 discussed the use of signal processing-based techniques such as classical distance-based tests, anomaly detection and statistics, and machine learning methods to identify faults in an array. It was shown that traditional non-robust, distance-based, anomaly detection techniques such as the Euclidean and Mahalanobis distance, are not adequate for fault detection in PV arrays. This is because the current and voltage of PV arrays are constrained by their electrical configuration and a fault in one module affects the entire array. Because of this, robust statistical methods such as the MCD estimator show better results as outlier detectors. The use of machine learning methods, such as artificial neural networks and support vector machines, are discussed as well. If a large and complete training dataset is available, these methods may be implemented for fault detection and diagnosis.

Chapter 5 discussed the design rules for stand-alone and grid-connected PV arrays. This chapter discussed the need for a reconfigurable topology facility for improving the performance of the arrays. An overview of the existing reconfiguration methodologies is provided. The requirements for reconfiguration of topologies are discussed. A typical topology reconfiguration algorithm is presented.

Chapter 6 presented an overview of PV array monitoring system. The existing methods of monitoring are discussed. The design considerations of a monitoring system that include the location and placement of sensors, the type of communication system to be employed, and the algorithms used for processing the data collected from the array are explained. The requirements of a GUI for PV system are discussed.

The decreasing cost of monitoring and control technology has created an opportunity to update PV array management practices. Automated systems must be developed to detect and respond to faults with little or no human involvement. Automated fault detection has the potential to dramatically reduce the mean time to repair of PV arrays, in some cases from weeks to less than a day. This increases array availability, and therefore power output. The addition of control circuitry to a PV array also allows dynamic reconfiguration of the array topology, either to mitigate a fault condition or to optimize the array size to provide a better yield.

APPENDIX A

Matlab Code

A.1 SANDIA PERFORMANCE MODEL

A.1.1 PARAMETERS FOR SHARP NT175U1

```
function [modelParams arrayParams envParams] = get_params()
clear modelParams

modelParams.name = 'Sharp NT-175U1'; modelParams.vintage = 2007;
modelParams.area = 1.3; modelParams.material = 'c-Si';
modelParams.series_cells = 72; modelParams.parallel_strings = 1;
modelParams.Isco = 5.40; modelParams.Voco = 44.4;
modelParams.Impo = 4.95; modelParams.Vmpo = 35.4;
modelParams.aIsc = .000351; modelParams.aImp = -.000336;
modelParams.C0 = 1.003; modelParams.C1 = -.003;
modelParams.BVoco = -.151; modelParams.mBVoc = 0;
modelParams.BVmpo = -.158; modelParams.mBVmp = 0;
modelParams.n = 1.323; modelParams.C2 = .001;
modelParams.C3 = -8.711;
modelParams.A = [.931498305 .059748475 -.010672586 .000798468
  -2.23567E-5];    -3.37053E-5];
modelParams.B = [1 -.002438 .0003103 -1.246E-5 2.112E-7 -1.359E-9];
modelParams.dTc = 3; modelParams.fd = 1;
modelParams.a = -3.56; modelParams.b = -.075;
modelParams.C4 = .992; modelParams.C5 = .008;
modelParams.Ixo = 5.32; modelParams.Ixxo = 3.51;
modelParams.C6 = 1.128; modelParams.C7 = -.128;
modelParams.e0 = 1000; modelParams.To = 25;


envParams.airmass = 1; envParams.aoi = 0;
envParams.T_ambient = 25; envParams.T_cell = 30;
envParams.irradiance = 1000; envParams.P_diffuse = 0;
```

```
arrayParams.trackType = 2;

arrayParams.tiltDirection = 180;

arrayParams.tiltAngle = 15;

arrayParams.size = 72; arrayParams.moduleType = 'Sharp NT-175U1';

arrayParams.maxVoltage = 600; arrayParams.minVoltage = 300;
arrayParams.maxCurrent = 55; arrayParams.tempCoeffAmpacity = -.85/100;

arrayParams.nSer = 12; arrayParams.nPar = 6;
arrayParams.resistance = 1;

arrayParams.location.longitude = -111.912186;
arrayParams.location.latitude = 33.423745;
arrayParams.location.altitude = 305;

arrayParams.utcOffset = -7;
```

A.1.2   IV CURVE FOR SANDIA PERFORMANCE MODEL

```
function [V I] = get_IV_curve(env, model) k = 1.38066E-23;
q = 1.602E-19;

f_airmass = max(0,polyval(fliplr(model.A),env.airmass));

f_aoi = polyval(fliplr(model.B),env.aoi);

delta_T = env.T_cell - model.To;

Isc = model.Isco * f_airmass *
  (f_aoi * env.irradiance + model.fd * env.P_diffuse)/model.e0
  * (1 + model.aIsc*delta_T);

Ee = Isc / (model.Isco * (1 + model.aIsc*delta_T));

Imp = model.Impo * (model.C0*Ee + model.C1 * Ee^2)
```

```
   *(1 + model.aImp*delta_T);

Vt = model.n * k * (env.T_cell + 273.15)/q;

BVoc = model.BVoco + model.mBVoc*(1- Ee);
Voc = max(0,model.Voco + model.series_cells*Vt*log(Ee)   + BVoc * delta_T);

BVmp = model.BVmpo + model.mBVmp*(1- Ee);
Vmp = max(0,model.Vmpo + model.C2*model.series_cells*Vt*log(Ee)
  + model.C3*model.series_cells*(Vt * log(Ee))^2 + BVmp * delta_T);

Ix = model.Ixo * (model.C4 * Ee + model.C5 * Ee^2)*
  (1 + model.aIsc*delta_T);
Ixx = model.Ixxo * (model.C6 * Ee + model.C7*Ee^2)
  *(1+ model.aImp*delta_T);

V = [0 Voc/2 Vmp (Voc + Vmp)/2 Voc]; I = [Isc Ix Imp Ixx 0];
if V(2) > V(3)      V = [0 Vmp (Voc + Vmp)/2 Voc];
    I = [Isc Imp Ixx 0]; end
if Voc == 0     V = 0;
    I = 0; end
```

Bibliography

[1] P. Wong, "Solar photovoltaic cell/module manufacturing activities 2009," US Energy Information Administration, Tech. Rep., Jan. 2011. Cited on page(s) 1

[2] J. Conti and P. Holtberg, "Levelized cost of new generation resources in the annual energy outlook 2011," US Energy Information Administration, Tech. Rep., Dec. 2010. Cited on page(s) 1

[3] D. King, J. Kratochvil, and W. Boyson, "Photovoltaic array performance model," Sandia National Laboratory, Tech. Rep., 2004. Cited on page(s) 2, 14, 15, 57, 61, 63

[4] A. Maish, C. Atcitty, S. Hester, D. Greenberg, D. Osborn, D. Collier, and M. Brine, "Photovoltaic system reliability," in *Photovoltaic Specialists Conference, 1997., Conference Record of the Twenty-Sixth IEEE*, Sep. 1997, pp. 1049–1054. Cited on page(s) 2

[5] K. Otani, T. Takashima, and K. Kurokawa, "Performance and reliability of 1 MW photovoltaic power facilities in AIST," in *Photovoltaic Energy Conversion, Conference Record of the 2006 IEEE 4th World Conference on*, vol. 2, May 2006, pp. 2046–2049.
DOI: 10.1109/WCPEC.2006.279904 Cited on page(s) 2

[6] W. De Soto, S. Klein, and W. Beckman, "Improvement and validation of a model for photovoltaic array performance," *Solar Energy*, vol. 80, no. 1, pp. 78–88, Aug. 2006.
DOI: 10.1016/j.solener.2005.06.010 Cited on page(s) 2, 14, 17

[7] T. Markvart, *Solar Electricity*, 2nd ed. Wiley, May 2000. Cited on page(s) 5, 7, 10

[8] E. Noll, *Wind/solar energy for radiocommunications and low-power electrical systems*. H.W. Sams, 1981. Cited on page(s) 6, 7, 13

[9] R. L. Boylestad and L. Nashelsky, *Electronic Devices and Circuit Theory*. Prentice-Hall of India Private Limited, 2002. Cited on page(s) 7

[10] J. Nelson, *The Physics of Solar Cells*. Imperial College Press, 2003. Cited on page(s) 7, 9, 11

[11] C. Honsberg and S. Bowden, *PV CDROM*. [Online]. Available: http://www.pveducation.org/pvcdrom Cited on page(s) 7, 8, 9, 10, 11, 12

[12] B. Duffie, *Solar Engineering of Thermal Processes*. Wiley, 1991. Cited on page(s) 7

[13] A. Jain and A. Kapoor, "A new method to determine the diode ideality factor of real solar cell using lambert w-function," *Solar Energy Materials and Solar Cells*, vol. 85, no. 3, pp. 391–396, 2005. DOI: 10.1016/j.solmat.2004.05.022 Cited on page(s) 9

[14] H. Bayhan and M. Bayhan, "A simple approach to determine the solar cell diode ideality factor under illumination," *Solar Energy*, vol. 85, no. 5, pp. 769–775, 2011. DOI: 10.1016/j.solener.2011.01.009 Cited on page(s) 9

[15] W. D. De Soto, "Improvement and validation of a model for photovoltaic array performance," Master's thesis, University of Wisconsin-Madison, 2004. DOI: 10.1016/j.solener.2005.06.010 Cited on page(s) 14, 16, 17

[16] B. Marion, J. Adelstein, K. Boyle, B. H. H. Hayden, T. Fletcher, B. Canada, D. Narang, D. Shugar, H. Wenger, A. Kimber, L. Mitchell, G. Rich, and T. Townsend, "Performance parameters for grid-connected pv systems," in *31st IEEE Photovoltaics Specialists Conference and Exhibition*, 2005. Cited on page(s) 15

[17] G. T. Klise and J. S. Stein, "Models used to assess the performance of photovoltaic systems," Sandia National Laboratories, Tech. Rep., 2009. Cited on page(s) 15, 16, 17

[18] "EES solver to find the reference parameters of five parameter model." [Online]. Available: http://sel.me.wisc.edu/software.shtml Cited on page(s) 17

[19] D. Picault, B. Raison, S. Bacha, J. Aguilera, and J. De La Casa, "Changing photovoltaic array interconnections to reduce mismatch losses: a case study," *International Conference on Environment and Electrical Engineering*, 2010. DOI: 10.1109/EEEIC.2010.5490027 Cited on page(s) 19

[20] H. Patel and V. Agarwal, "MATLAB-Based modeling to study the effects of partial shading on pv array characteristics," *Energy Conversion, IEEE Transactions on*, vol. 23, no. 1, pp. 302–310, Mar. 2008. DOI: 10.1109/TEC.2007.914308 Cited on page(s) 23

[21] D. Nguyen and B. Lehman, "Modeling and simulation of solar PV arrays under changing illumination conditions," in *Computers in Power Electronics, 2006. COMPEL '06. IEEE Workshops on*, July 2006, pp. 295–299. DOI: 10.1109/COMPEL.2006.305629 Cited on page(s) 23

[22] V. Quaschning and R. Hanitsch, "Numerical simulation of current-voltage characteristics of photovoltaic systems with shaded solar cells," *Solar Energy*, vol. 56, no. 6, 1996. DOI: 10.1016/0038-092X(96)00006-0 Cited on page(s) 23

[23] F. Spertino and J. Akilimali, "Are manufacturing I-V mismatch and reverse currents key factors in large photovoltaic arrays?" *Industrial Electronics, IEEE Transactions on*, vol. 56, no. 11, pp. 4520–4531, Nov. 2009. DOI: 10.1109/TIE.2009.2025712 Cited on page(s) 27, 28

[24] R. Hammond, D. Srinivasan, A. Harris, K. Whitfield, and J. Wohlgemuth, "Effects of soiling on PV module and radiometer performance," in *Photovoltaic Specialists Conference, 1997., Conference Record of the Twenty-Sixth IEEE*, 29 1997, pp. 1121–1124. DOI: 10.1109/PVSC.1997.654285 Cited on page(s) 28

[25] R. C. Quick, "Ground fault circuit interrupter—design and operating characteristics," *Industry Applications, IEEE Transactions on*, vol. IA-11, no. 1, pp. 50–55, 1975. DOI: 10.1109/TIA.1975.349257 Cited on page(s) 31

[26] J. Wiles, "Ground-fault protection for PV systems," *IAEI News*, pp. 2–7, 2008. Cited on page(s) 31

[27] *NFPA 70: National Electrical Code*, NFPA Std., 2008. Cited on page(s) 31

[28] G. Gregory and G. Scott, "The arc-fault circuit interrupter: an emerging product," *Industry Applications, IEEE Transactions on*, vol. 34, no. 5, pp. 928–933, 1998. DOI: 10.1109/28.720431 Cited on page(s) 31

[29] G. Gregory, K. Wong, and R. Dvorak, "More about arc-fault circuit interrupters," *Industry Applications, IEEE Transactions on*, vol. 40, no. 4, pp. 1006–1011, 2004. DOI: 10.1109/IAS.2003.1257720 Cited on page(s) 31

[30] M. Naidu, T. Schoepf, and S. Gopalakrishnan, "Arc fault detection scheme for 42-V automotive dc networks using current shunt," *Power Electronics, IEEE Transactions on*, vol. 21, no. 3, pp. 633–639, May 2006. DOI: 10.1109/TPEL.2006.872385 Cited on page(s) 31

[31] H. Haeberlin and M. Kaempfer, "Measurement of damages at bypass diodes by induced voltages and currents in PV modules caused by nearby lightning currents with standard waveform," in *23rd European Photovoltaic Solar Energy Conference*, 2008. DOI: 10.4229/23rdEUPVSEC2008-4AV.3.54 Cited on page(s) 31

[32] K.-J. Tseng, Y. Wang, and D. Vilathgamuwa, "An experimentally verified hybrid cassie-mayr electric arc model for power electronics simulations," *Power Electronics, IEEE Transactions on*, vol. 12, no. 3, pp. 429–436, may 1997. DOI: 10.1109/63.575670 Cited on page(s) 32, 33

[33] A. M. Cassie, "Arc rupture and circuit serverity: A new theory," CIGRE, Tech. Rep., 1939. Cited on page(s) 32

[34] O. Mayr, "Beiträge zur theorie des statischen und des dynamischen lichtbogens," *Electrical Engineering (Archiv fur Elektrotechnik)*, vol. 37, pp. 588–608, 1943, 10.1007/BF02084317. [Online]. Available: http://dx.doi.org/10.1007/BF02084317 DOI: 10.1007/BF02084317 Cited on page(s) 32

[35] G. Idarraga Ospina, D. Cubillos, and L. Ibanez, "Analysis of arcing fault models," in *Transmission and Distribution Conference and Exposition: Latin America, 2008 IEEE/PES*, aug. 2008, pp. 1–5. DOI: 10.1109/TDC-LA.2008.4641860 Cited on page(s) 33

[36] U. Habedank, "Application of a new arc model for the evaluation of short-circuit breaking tests," *Power Delivery, IEEE Transactions on*, vol. 8, no. 4, pp. 1921–1925, Oct. 1993. DOI: 10.1109/61.248303 Cited on page(s) 34

[37] N. Bosco, "Reliability concerns associated with PV technologies," National Renewable Energy Laboratory, Tech. Rep., Apr. 2010. Cited on page(s) 34

[38] F. Chan and H. Calleja, "Reliability: A new approach in design of inverters for PV systems," in *International Power Electronics Congress, 10th IEEE*, oct. 2006, pp. 1–6. DOI: 10.1109/CIEP.2006.312159 Cited on page(s) 35

[39] S. Vergura, G. Acciani, V. Amoruso, and G. Patrono, "Inferential statistics for monitoring and fault forecasting of PV plants," in *ISIE 2008. Industrial Electronics, 2008, IEEE International Symposium on*, Jun. 2008, pp. 2414–2419. DOI: 10.1109/ISIE.2008.4677264 Cited on page(s) 37

[40] T. Takashima, J. Yamaguchi, K. Otani, K. Kato, and M. Ishida, "Experimental studies of failure detection methods in PV module strings," in *Photovoltaic Energy Conversion, Conference Record of the 2006 IEEE 4th World Conference on*, vol. 2, 7–12 2006, pp. 2227–2230. DOI: 10.1109/WCPEC.2006.279952 Cited on page(s) 37

[41] C. Griesbach, "Fault-tolerant solar array control using digital signal processing for peak power tracking," in *Energy Conversion Engineering Conference, 1996. IECEC 96. Proceedings of the 31st Intersociety*, vol. 1, 11–16 1996, pp. 260–265 vol.1. DOI: 10.1109/IECEC.1996.552881 Cited on page(s) 37

[42] A. Drews, A. De Keizer, H. Beyer, E. Lorenz, J. Betcke, W. van Sark, W. Heydenreich, E. Wiemken, S. Stettler, P. Toggweiler *et al.*, "Monitoring and remote failure detection of grid-connected PV systems based on satellite observations," *Solar energy*, vol. 81, no. 4, pp. 548–564, 2007. DOI: 10.1016/j.solener.2006.06.019 Cited on page(s) 37

[43] D. Stellbogen, "Use of PV circuit simulation for fault detection in PV array fields," in *Photovoltaic Specialists Conference, 1993., Conference Record of the Twenty Third IEEE*, May 1993, pp. 1302–1307. DOI: 10.1109/PVSC.1993.346931 Cited on page(s) 37

[44] H. Zhiqiang and G. Li, "Research and implementation of microcomputer online fault detection of solar array," in *Computer Science Education, 2009. ICCSE '09. 4th International Conference on*, Jul. 2009, pp. 1052–1055. DOI: 10.1109/ICCSE.2009.5228541 Cited on page(s) 37, 59

[45] P. J. Rousseeuw and K. v. Driessen, "A fast algorithm for the minimum covariance determinant estimator," *Technometrics*, vol. 41, no. 3, pp. 212–223, Aug. 1999. DOI: 10.2307/1270566 Cited on page(s) 39, 40

[46] V. Hautamaki, I. Karkkainen, and P. Franti, "Outlier detection using k-nearest neighbour graph," in *Pattern Recognition, 2004. ICPR 2004. Proceedings of the 17th International Conference on*, vol. 3, Aug. 2004, pp. 430–433. DOI: 10.1109/ICPR.2004.1334558 Cited on page(s) 39

[47] F. Angiulli and C. Pizzuti, "Fast outlier detection in high dimensional spaces," in *Principles of Data Mining and Knowledge Discovery*, ser. Lecture Notes in Computer Science, T. Elomaa, H. Mannila, and H. Toivonen, Eds. Springer Berlin / Heidelberg, 2002, vol. 2431, pp. 43–78. DOI: 10.1007/3-540-45681-3_2 Cited on page(s) 39

[48] P. Rousseeuw, "Multivariate estimation with high breakdown point," *Mathematical statistics and applications*, vol. 8, pp. 283–297, 1985. DOI: 10.1007/978-94-009-5438-0_20 Cited on page(s) 40

[49] S. Dreiseitl and L. Ohno-Machado, "Logistic regression and artificial neural network classification models: a methodology review," *Journal of Biomedical Informatics*, vol. 35, no. 5–6, pp. 352–359, 2002. DOI: 10.1016/S1532-0464(03)00034-0 Cited on page(s) 40

[50] S. Haykin, *Neural Networks and Learning Machines*, 3rd ed. Prentice Hall, 2009. Cited on page(s) 41

[51] A. Al-Amoudi and L. Zhang, "Application of radial basis function networks for solar-array modelling and maximum power-point prediction," *Generation, Transmission and Distribution, IEEE Proceedings*, vol. 147, no. 5, pp. 310–316, Sep. 2000. DOI: 10.1049/ip-gtd:20000605 Cited on page(s) 42

[52] C. Cortes and V. Vapnik, "Support-vector networks," *Machine Learning*, vol. 20, pp. 273–297, 1995, 10.1007/BF00994018. DOI: 10.1023/A:1022627411411 Cited on page(s) 44

[53] M. V. Paukshto and K. Lovetskiy, "Invariance of single diode equation and its application," in *Photovoltaic Specialists Conference, 2008. PVSC '08. 33rd IEEE*, 11–16 May 2008, pp. 1–4. DOI: 10.1109/PVSC.2008.4922501 Cited on page(s) 47

[54] V. Badescu, "Simple optimization procedure for silicon-based solar cell interconnection in a series-parallel PV module," *Energy Conversion and Management*, vol. 47, no. 9–10, pp. 1146–1158, 2006. DOI: 10.1016/j.enconman.2005.06.018 Cited on page(s) 47

[55] "IEEE guide for array and battery sizing in stand-alone photovoltaic (PV) systems," *IEEE Std 1562-2007*, pp. i–22, 12 2008. DOI: 10.1109/IEEESTD.2008.4518937 Cited on page(s) 47

[56] G. S. E. S. DGS, *Planning and Installing Photovoltaic Systems : A Guide for Installers, Architects and Engineers*, 2nd ed. Earthscan Ltd, 2007. Cited on page(s) 48

[57] D. Picault, B. Raison, S. Bacha, J. de la Casa, and J. Aguilera, "Forecasting photovoltaic array power production subject to mismatch losses," *Solar Energy*, vol. 84, no. 7, pp. 1301–1309, 2010. DOI: 10.1016/j.solener.2010.04.009 Cited on page(s) 48

[58] G. Velasco, J. Negroni, F. Guinjoan, and R. Pique, "Irradiance equalization method for output power optimization in plant oriented grid-connected PV generators," in *Power Electronics and Applications, 2005 European Conference on*, 2005, p. 10. DOI: 10.1109/EPE.2005.219300 Cited on page(s) 52

[59] D. Nguyen, "Modeling and reconfiguration of solar photovoltaic arrays under non-uniform shadow conditions," Master's thesis, Northeastern University, 2008. Cited on page(s) 53

[60] D. Nguyen and B. Lehman, "A reconfigurable solar photovoltaic array under shadow conditions," in *Applied Power Electronics Conference and Exposition, 2008. APEC 2008. Twenty-Third Annual IEEE*, 24–28 2008, pp. 980–986. DOI: 10.1109/APEC.2008.4522840 Cited on page(s) 53

[61] L. Tria, M. Escoto, and C. Odulio, "Photovoltaic array reconfiguration for maximum power transfer," in *TENCON 2009 - 2009 IEEE Region 10 Conference*, Jan. 2009, pp. 1–6. DOI: 10.1109/TENCON.2009.5395965 Cited on page(s) 54

[62] C. Craig Jr, "Solar array switching unit," US Patent 6 060 790, May 9, 2000. Cited on page(s) 54

[63] "IEEE standard for interconnecting distributed resources with electric power systems," *IEEE Std 1547–2003*, pp. 1–16, 2003. DOI: 10.1109/IEEESTD.2003.94285 Cited on page(s) 57, 58

[64] H. Braun, S. T. Buddha, V. Krishnan, C. Tepedelenlioglu, A. Spanias, T. Yeider, and T. Takehara, "Signal processing for fault detection in photovoltaic arrays," in *IEEE Int. Conf. on Acoustics, Speech, and Signal Processing*, 2012. Cited on page(s) 59

[65] E. Dirks, A. Gole, and T. Molinski, "Performance evaluation of a building integrated photovoltaic array using an internet based monitoring system," in *Power Engineering Society General Meeting, 2006. IEEE*, 2006. DOI: 10.1109/PES.2006.1709233 Cited on page(s) 59

[66] W. Kolodenny, M. Prorok, T. Zdanowicz, N. Pearsall, and R. Gottschalg, "Applying modern informatics technologies to monitoring photovoltaic (PV) modules and systems," in *Photovoltaic Specialists Conference, 2008. PVSC '08. 33rd IEEE*, May 2008, pp. 1–5. DOI: 10.1109/PVSC.2008.4922829 Cited on page(s) 60

[67] M. Zahran, Y. Atia, A. Al-Hussain, and I. El-Sayed, "LabVIEW based monitoring system applied for PV power station," in *Proceedings of the 12th WSEAS international conference on Automatic control, modelling & simulation*, ser. ACMOS'10. Stevens Point, Wisconsin, USA:

World Scientific and Engineering Academy and Society (WSEAS), 2010, pp. 65–70. Cited on page(s) 61

[68] X. Carcelle, *Power Line Communications in Practice*, 1st ed.   Artech House Publishers, 2009. Cited on page(s) 64

[69] H. Ferreira, H. Grove, O. Hooijen, and A. Han Vinck, "Power line communications: an overview," in *AFRICON, 1996., IEEE AFRICON 4th*, vol. 2, 24–27 1996, pp. 558–563 vol. 2. DOI: 10.1109/MP.2004.1343222 Cited on page(s) 64

[70] L. Surhone, M. Timpledon, and S. Marseken, *RS-232: Telecommunication, Data Terminal Equipment, Data Circuit-Terminating Equipment, Serial Port, ITU-T, Electronic Industries Alliance, Asynchronous Serial Communication.*   Betascript Publishers, 2010. Cited on page(s) 64

[71] D. Gislason, *Zigbee wireless networking*, ser. Safari Books Online.   Elsevier, 2008. Cited on page(s) 65

[72] J.-S. Lee, Y.-W. Su, and C.-C. Shen, "A comparative study of wireless protocols: Bluetooth, UWB, ZigBee, and Wi-Fi," in *Industrial Electronics Society, 2007. IECON 2007. 33rd Annual Conference of the IEEE*, nov. 2007, pp. 46–51. DOI: 10.1109/IECON.2007.4460126 Cited on page(s) 65

[73] N. Baker, "Zigbee and Bluetooth strengths and weaknesses for industrial applications," *Computing Control Engineering Journal*, vol. 16, no. 2, pp. 20–25, april-may 2005. DOI: 10.1049/ccej:20050204 Cited on page(s) 65

Authors' Biographies

MAHESH BANAVAR

Mahesh K. Banavar is a post-doctoral researcher in the School of Electrical, Computer and Energy Engineering at Arizona State University. He received a B.E. degree in Telecommunications Engineering from Visvesvaraya Technological University, Karnataka, India in 2005, and M.S. and Ph.D. degrees, both in Electrical Engineering, from Arizona State University in 2007 and 2010, respectively. His research area is Signal Processing and Communications, with specific interest in Wireless Communications, Sensor Networks, Distributed Inference, Localization, and applications of statistical signal processing. He is a member of MENSA and the Eta Kappa Nu honor society.

HENRY BRAUN

Henry Braun completed his B.S. and M.S. degrees in Electrical Engineering at Arizona State University in Tempe, Arizona, and is continuing as a Ph.D. student. During an internship at the NASA Jet Propulsion Lab working in the computer vision field, he developed a strong interest in information processing and decision making. This interest is reflected in his recent work on automated fault detection in photovoltaic arrays. Other research interest areas include compressive sensing and automatic target recognition and tracking. In addition to NASA, Henry's research sponsors have included Paceco Corp. and Raytheon missile systems.

SANTOSHI TEJASRI BUDDHA

Santoshi Tejasri Buddha received her M.S. degree in signal processing and communications field from Arizona State University in November, 2011. She did her Master's thesis on the topic "Topology Reconfiguration to Improve the Photovoltaic Array Performance" under the guidance of Drs. Andreas Spanias and Cihan Tepedelenlioglu. She currently works as a Digital Signal Processing engineer at Advanced Bionics.

VENKATACHALAM KRISHNAN

Venkatachalam Krishnan graduated with an M.S. degree in Electrical Engineering from Arizona State University in 2012. His Master's thesis focused on sensor placement and GUI for PV array monitoring systems. Prior to his M.S., he was a software engineer for Analog Devices working on video codecs. Currently, he works as a wireless firmware engineer in Qualcomm.

ANDREAS SPANIAS

Andreas Spanias is a Professor in the School of Electrical, Computer, and Energy Engineering at Arizona State University. He is also the founder and director of the SenSIP industry consortium. His research interests are in the areas of adaptive signal processing, speech processing, and audio sensing. He and his student team developed the computer simulation software Java-DSP (J-DSP - ISBN 0-9724984-0-0). He is the author of two text books: *Audio Processing and Coding* by Wiley and *DSP: An Interactive Approach.* He served as Associate Editor of the IEEE Transactions on Signal Processing and as General Co-chair of IEEE ICASSP-99. He also served as the IEEE Signal Processing Vice-President for Conferences. Andreas Spanias is co-recipient of the 2002 IEEE Donald G. Fink paper prize award and was elected Fellow of the IEEE in 2003. He served as Distinguished lecturer for the IEEE Signal Processing Society in 2004.

TORU TAKEHARA

Toru Takehara, Ph.D., is a researcher at Mitsui Engineering & Shipbuilding, Co., Ltd. Japan. He was President and CEO of PACECO CORP. in the U.S. for 10 years. After graduating from a Mechanical Engineering course of Osaka University in Japan (1974), he worked as crane design Engineer, developing container AGVs and other automation equipment. He is the inventor of the PV Module Monitoring System.

CIHAN TEPEDELENLIOGLU

Cihan Tepedelenlioglu (S'97-M'01) was born in Ankara, Turkey in 1973. He received his B.S. degree with highest honors from Florida Institute of Technology in 1995, and his M.S. degree from the University of Virginia in 1998, both in Electrical Engineering. From January 1999 to May 2001 he was a research assistant at the University of Minnesota, where he completed his Ph.D. degree in Electrical and Computer Engineering. He is currently an Associate Professor of Electrical Engineering at Arizona State University. He was awarded the NSF (early) Career grant in 2001, and has served as an Associate Editor for several IEEE Transactions including IEEE Transactions on Communications, and IEEE Signals Processing Letters. His research interests include statistical signal processing, system identification, wireless communications, estimation and equalization algorithms for wireless systems, multi-antenna communications, filter banks and multirate systems, OFDM, ultra-wideband systems, distributed detection and estimation, and data mining for PV systems.

TED YEIDER

Ted Yeider is the Business Development Manager, at Paceco Corp. He has been with Paceco for more than three years and has worked with research and development of prototype products in several industries. A veteran of the United States Marine Corps, educated as a marine engineer, Mr. Yeider has been involved with industrial engineering inspection work; marine engineering instruction; oil and gas tanker engineering operations and maintenance, oil refinery in-service and turnaround maintenance activities; marine container terminal equipment engineering and main-tenance. In his current role, Mr. Yeider works in the business development arena with a leading Marine Terminal Equipment Manufacturer. With a career that started over 30 years ago, Yeider has accumulated valuable experience and knows what it takes to maintain mission critical assets. Yeider's experience working with some of the top global companies has made him a knowledgeable resource for leaders looking to develop best practices and improve results. Yeider holds a B.S. degree in Marine Engineering and currently resides in Northern California.

SHINICHI TAKADA

Shinichi Takada is Chief Technology Officer, Senior executive of Applied Core Technology, Inc., Hayward, California. He works closely with institutes to develop renewable energy conservation technology such as an advanced Photo Voltaic solar generation. His own developing group is also developing high-intensity X-ray source for a medical imaging system. Until 1998, he worked at Seiko Instruments where he had vast experience and made great achievements with the Instrumentation of Analytical equipment such as Induction Coupled Plasma Spectrometer, Induction Coupled Plasma Mass Spectrometer, X-ray Fluorescent Analyzer, and trace gas in metal analyzer. He graduated from Tokyo University of Science in 1978 with an M.S. degree in Industrial Chemical Engineering.